Green Energy and Technology

Climate change, environmental impact and the limited natural resources urge scientific research and novel technical solutions. The monograph series Green Energy and Technology serves as a publishing platform for scientific and technological approaches to "green"—i.e. environmentally friendly and sustainable—technologies. While a focus lies on energy and power supply, it also covers "green" solutions in industrial engineering and engineering design. Green Energy and Technology addresses researchers, advanced students, technical consultants as well as decision makers in industries and politics. Hence, the level of presentation spans from instructional to highly technical.

Indexed in Scopus.

Indexed in Ei Compendex.

Yatindra Gopal · Dheeraj Kumar Dhaked ·
Kaibalya Prasad Panda · Julio C. Rosas-Caro
Editors

Multilevel Inverter Technology for Renewable Energy Systems

Enhancing Efficiency and Sustainability Through Advanced Power Conversion Solutions

Springer

Editors

Yatindra Gopal
Department of Electrical Engineering
National Institute of Technology
Arunachal Pradesh
Jote, Arunachal Pradesh, India

Kaibalya Prasad Panda
Department of Electrical Engineering
School of Energy Technology
Pandit Deendayal Energy University
Gandhinagar, Gujarat, India

Dheeraj Kumar Dhaked
Veermata Jijabai Technological Institute
Mumbai, Maharashtra, India

Julio C. Rosas-Caro
Facultad de Ingenieria
Universidad Panamericana
Zapopan, Jalisco, Mexico

ISSN 1865-3529 ISSN 1865-3537 (electronic)
Green Energy and Technology
ISBN 978-3-032-03588-2 ISBN 978-3-032-03589-9 (eBook)
https://doi.org/10.1007/978-3-032-03589-9

This Springer imprint is published by the registered company Springer Nature Switzerland AG
The registered company address is: Gewerbestrasse 11, 6330 Cham, Switzerland

Contents

Multilevel Inverter Topology with Reduced Components for Renewable Energy Applications 1
Dhananjay Kumar, Vishal Rathore, Kasinath Jena, Ramjee Prasad Gupta, Santosh Kumar Gupta, Vinod Kumar Mehta, Kumar Rahul, and Shahzad Ahsan

Multilevel VSI for Battery Powered Applications with Novel PWM Control Scheme 21
Shailesh Kumar Gupta

Evolution of Reduced-Component Switched-Capacitor Boost Inverters for Photovoltaic System 45
Sumant Kumar Dalai, Kaibalya Prasad Panda, and Gayadhar Panda

Analysis of Solar PV-Powered Reduced Component Switched-Capacitor-Based Multilevel Inverter 61
J. Vijaychandra, Yatindra Gopal, K. Veda Prakash, B. Vanajakshi, Lukasz Knypinski, D. Vijaya Kumar, and N. Sowjanya

High Voltage Gain Multilevel Inverter Topology with Reduced Device Count and Low Voltage Stress 75
Prayag Chandra Barik, Narayan Nayak, Sidharth Samantara, Ashwini Kumar Sahu, Kasinath Jena, Dhananjay Kumar, Vishal Rathore, and Ramjee Prasad Gupta

Novel Switched Capacitor Self-Balanced Reduced Switch Multilevel Inverter for Photovoltaic System 87
Yatindra Gopal, J. Vijaychandra, Akanksha Kumari, Kaibalya Prasad Panda, and B. Parusharamulu

Common-Ground Boosting Multilevel Inverter Topology with Optimal Components Count 107
Kaibalya Prasad Panda, Sumant Kumar Dalai, Yatindra Gopal, and Gayadhar Panda

Investigation of Fault-Tolerant Multilevel Inverter Topologies 121
Dhananjay Kumar, Santosh Kumar Gupta, Vishal Rathore, Kasinath Jena, Ramjee Prasad Gupta, Samar Anand, Kumar Rahul, and Shahzad Ahsan

Machine Learning Based Fault Detection and Classification in Multilevel Inverter for Industrial Applications 143
Niraj Kumar Dewangan, N. D. Jeevan, Vivek Gurjar, Kasinath Jena, and Dhananjay Kumar

Advancement in MMC Topologies and Balancing Techniques for Medium- and High-Power Applications 157
Jyoti Ranjan Dash, Rudranarayan Pradhan, and Pramod Agarwal

Index 177

Multilevel Inverter Topology with Reduced Components for Renewable Energy Applications

Dhananjay Kumar, Vishal Rathore, Kasinath Jena, Ramjee Prasad Gupta, Santosh Kumar Gupta, Vinod Kumar Mehta, Kumar Rahul, and Shahzad Ahsan

Abstract This work presents a reduced device count (RDC) asymmetrical cross-connected seventeen-level T-type (CT-type) inverter. The proposed approach includes eight power switches, four self-voltage balancing capacitors and two direct current (DC) sources. The suggested architecture eliminates the need for an H-bridge to generate negative half-cycle, hence reducing the voltage stress across the switches. The capacitor and DC sources cascaded with the switches to produce required number of levels. This design is exceptionally conducive to the incorporation of the renewable energy sources into the grid because of its significant self and high voltage balancing capability. Gate driver signals for the inverter circuit are produce by pulse width modulation based on the level-shifted technique. Comparative analysis is undertaken to validate the efficacy of the suggested topology which is distinguished against previously reported topologies based on the number of DC sources and capacitor count, switches, drivers and total standing voltage. Finally, the feasibility of the proposed topology is examined by the simulated results and is further verified by a laboratory prototype.

Keywords Pulse width modulation · CT-type · Total harmonic distortion · Reduced device count · Multilevel inverter · Total standing voltage

---

D. Kumar (✉) · S. K. Gupta · V. K. Mehta
Department of Electrical Engineering, Government Engineering College Siwan, Bihar, India
e-mail: dhananjaymanit23@gmail.com

V. Rathore
Department of Electrical Engineering, Maulana Azad National Institute of Technology, Bhopal, Madhya Pradesh, India

K. Jena
Department of Electrical and Electronic Engineering, ARKA JAIN University, Jamshedpur, India

R. P. Gupta · S. Ahsan
Department of Electrical Engineering, MIT Muzaffarpur, Bihar, India

K. Rahul
Department of Interdisciplinary Sciences, NIFTEM, Kundli, Sonepat, India

Y. Gopal et al. (eds.), *Multilevel Inverter Technology for Renewable Energy Systems*, Green Energy and Technology, https://doi.org/10.1007/978-3-032-03589-9_1

1 Introduction

MLIs have played a leading role in various applications for high and medium voltage/ power ratings. The MLIs usage in large areas are dominated by power electronics devices like electric vehicles, wind turbines, active power filters, grid-connected PV systems, induction motor control etc., because of their ability to synthesize high-voltage outputs [1–5]. MLIs can generate various output voltage levels by incorporating multiple capacitors, power semiconductor switches, DC supplies, and diodes. Technically, MLIs are sound in delivering high efficiency, modularity, low THD, high voltage levels, etc. MLIs are broadly classified into three types: Neutral Point Clamped (NPC), Cascaded H-bridge (CHB), and Flying Capacitor (FC). CHB has gained more interest among researchers out of these technologies as it overcomes the drawback of the other two, i.e., high switching losses, unbalanced DC voltage, and the use of more capacitors. Some recently developed and traditional MLI topologies presented in [6, 7] have been briefly reviewed for this work.

Multiple DC source-based CHBs are more appealing than single DC MLIs as of their higher reliability and modularity. Moreover, asymmetric CHB has less complex control than symmetric configurations and significantly increases voltage levels with reduced device count [8, 9]. On the other hand, the count of isolated DC voltage sources required in CHB is significantly higher. Previously, MLIs were limited by needing more DC sources and semiconductor switches. Over Multiple DC source-based CHBs are more appealing than single DC MLIs as of their higher reliability and modularity. Moreover, asymmetric CHB has less complex control than symmetric configurations and significantly increases voltage levels with reduced device count [8, 9]. On the other hand, the count of isolated DC voltage sources required in CHB is significantly higher. Previously, MLIs were limited by needing more DC sources and semiconductor switches. Over the last few years, extensive research has been conducted on improving MLI configuration in every feasible way. In [10–12], MLIs with switched-DC, switched-diode, and switched-capacitor configurations have been proposed. In [13], asymmetric switched-DCMLI has lesser components than traditional circuitry. The proposed topologies in [14, 15] are cost-effective and perform better in the switched DC source. Even with a reduced device count, these configurations can provide both negative and positive voltage polarity. Besides that, significantly higher voltage levels can be obtained by connecting multiple fundamental units in series. In the presence of an integrated H-bridge, optimized structures in [16, 17] revealed that 15 levels could be generated using 16 interrupter numbers and 7 DC sources in the basic unit. Voltage levels using different DC sources can be generated by having lower switching stress by modifying the same. References [18–20] presented and validated the symmetrical and asymmetrical topologies using various pulse width modulation techniques (PWM). References [21–23] investigated switched DC-based topologies that can replace traditional MLIs. Two switches connected by a DC link to generate the required voltage level are proposed in [24]. A negative voltage level was created by incorporating an H-bridge into the initial design. This MLI modification was improved in [25], which added two capacitors to get more levels out of each module.

Conventional MLIs can be combined to produce multiple voltage levels in a hybrid form. Some hybrid topologies proposed in [26] are classified as symmetrical MLIs. Because of the unequal DC input voltage, symmetric MLIs generate high voltage levels with reduced device count (RDC). An unconventional asymmetrical topology is presented in [27] by utilizing power switches to demonstrate higher voltage levels, reducing stress from semiconductor switches. Also, a modified H-bridge topology proposed in [28] has unequal voltage DC supplies. Therefore, different bridges in symmetrical MLIs operate at different voltages. Consequently, switches associated with the high voltage bridge must be capable of withstanding high voltage. However, the primary design objective of MLIs may be compromised because this method uses low-rating switches for high-power applications, even though it can significantly increase voltage levels. This disadvantage affects asymmetrical MLIs that use the voltage balancing method. Furthermore, the asymmetrical MLIs proposed in [26–28] are composed of self-balancing bridges with equal DC-link voltages primarily designed in a symmetrical configuration. So, no self-balancing circuit is required, making inverter control simpler.

The proposed MLI is intended to increase the output voltage level while ensuring the smart setup of semiconductor switches. The economical design also ensures that the number of power devices is minimal. This paper introduces asymmetrical CT-type MLI configurations. In a symmetrical configuration, there are two DC voltage supplies and four capacitors of the same rating, allowing it to produce four negatives, four positives, and one zero for nine different voltage levels. DC voltage sources in CT-type modules have followed the ratio of 1:3 for asymmetric configuration. As a result, the CT-type module can produce 17 (eight negatives, eight positives, and zero) distinct voltage levels. CT-type modules are included in the suggested asymmetric topology so as to produce that voltage levels of the negative half without the additional assistance of a CHB. Only 2 DC sources, 4 capacitors, and 8 switches were used.

This paper has the following structure: Sect. 2 discusses the circuit and the working of the proposed CT-Type-MLI. In addition, this section also provides asymmetric configuration and modes of operation, as well as module extension with output voltage level generation. Section 3 presents the modulation and control techniques followed by the simulation results for the proposed CT-Type MLI. The complete test rig and laboratory results are discussed in Sect. 4. The proposed MLI's power loss and efficiency are discussed thoroughly in Sect. 5 followed by the detailed comparative analysis of the recent topologies with the proposed CT-Type topology presented in Sect. 6. Finally, Sect. 7 contains the conclusion of the paper.

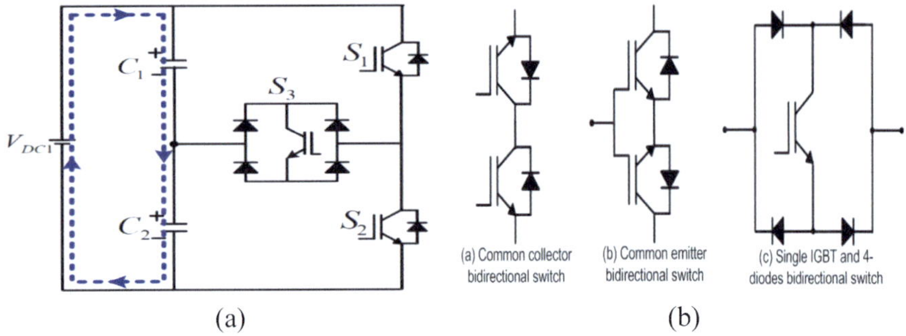

Fig. 1 **a** Basic cell of single-phase CT-type multilevel inverter, **b** Various arrangements for bidirectional switches

2 Generalised Topology of the Proposed MLI

2.1 Circuit Configuration and Its Working

Figure 1a illustrates a schematic diagram of a basic CT-type multilevel inverter. It consists of 1 DC source, 2 capacitors, 2 unidirectional switches, and 1 bidirectional switch. The DC supply's bidirectional switches prevent short circuit currents by facilitating blockage of both current polarities. Moreover, using the proposed MLI DC-Link capacitors inherently achieves self-voltage balance. This module has generated the three-level output voltage. Various arrangements for bi-directional switches have been shown in Fig. 1b.

Figure 2 illustrates the generalized structure of the proposed asymmetric single-phase CT-Type MLI module. A pair of power switches connect the two fundamental modules, L and R. The L module and R module are connected by unidirectional switches S_1 and S_2, and vice versa. Both modules are connected to the load (L and R). This arrangement can produce multiple voltage levels despite having a reduced power device count. Higher voltage levels are expected to significantly lower total harmonic distortion (THD) and low switching frequency.

2.2 Analytical Analysis of Capacitors Values

The capacitor starts charging and discharges as connected series with the load and in parallel across the voltage source respectively [7]. In the proposed CT-type MLI, the capacitor's charging-discharging significantly impacts levels of the output voltage. Therefore, it is critical to identify the appropriate value of the capacitor. The capacitance value is determined by the operating frequency and the load current. The capacitor's largest voltage ripple occurs during the discharging phase, as described in [1]. Consequently, the capacitance value is calculated using the worst case of the

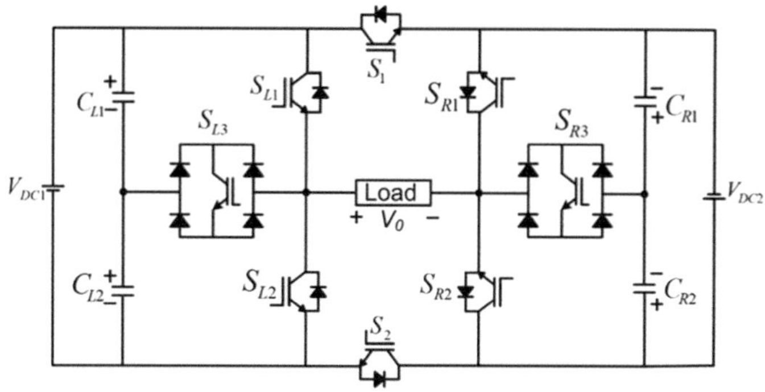

Fig. 2 Generalized single-phase CT-Type MLI module

capacitor's discharge duration. The voltage ripple must be within a specific range to be considered acceptable. When the ripple voltage is kept to a minimum, it reduces losses, thereby improving the efficiency of the proposed topology. In Fig. 2, it can be seen that the DC input voltage source is connected in series with both capacitors. As a result, the identical current flows through them during charging so that the same charge accumulates on them. Therefore,

$$I_{C1} = I_{C2} \;\; \Rightarrow Q_{C1} = Q_{C2} \;\; \Rightarrow C_1.V_{C1} = C_2 V_{C2} \tag{1}$$

When identical capacitance values are used, Eq. (1) achieves

$$\begin{cases} C_1 V_{C1} = C_2 V_{C2} \\ V_{C1} = V_{C2} \end{cases} \Rightarrow \;\; C_1 = C_2 \tag{2}$$

As a result, Eq. (2) provides identical values for the capacitances C_1 and C_2, i.e., $C_1 = C_2$. Each capacitor's maximum charge drawn is expressed as:

$$\Delta Q_C = I_L \int_{t_S}^{t_f} \sin(2\pi f_0 t)\, dt \tag{3}$$

where, time interval $[t_S, t_f]$ demonstrates the worst conceivable state for each capacitor. I_L and f_0 indicate the load current amplitude and output fundamental frequency, respectively. Finally, it should be noticed that the capacitance values of capacitors C_1 and C_2 are identical, when load connected to the equivalent capacitance (C_{eq}) is expressed as:

$$C_{eq.} \geq \frac{\Delta Q_C}{K.V_{eq}} \tag{4}$$

The output current for the resistive load is given by the equation as:

$$I_L(t) = \begin{Bmatrix} \frac{V_{dc}}{R_L} & for \quad t_1 \le t \le t_2 \\ \frac{2V_{dc}}{R_L} & for \quad t_2 \le t \le t_3 \\ \frac{3V_{dc}}{R_L} & for \quad t_3 \le t \le \frac{T}{4} \end{Bmatrix} \tag{5}$$

However, the output current for the resistive-inductive load is given by the equation as:

$$I_L(t) = I_{\max} \sin(\omega t - \alpha)\, dt \tag{6}$$

The appropriate capacitor ranges under *RL,* and pure resistive loads are obtained from (7).

$$C_1 = C_2 \ge \frac{2 \int_{t_3}^{T/4} I_L(t)dt}{K.V_{eq}} \tag{7}$$

The proposed CT-type topology's circuit parameters are R = 20 and 40, L = 20 and 40, k = 0.05, f_{ref} = 50 Hz, and the equivalent capacitance of both capacitors is connected in series is $C_1 = C_2 \ge 1780\,\mu F$. Additionally, it indicates that we must choose at least two identical capacitors for the inverter with the aforementioned specifications, the load, and the 5% ripple capacitor voltage tolerance. Therefore, in Sects. 3 and 4, the closest widely known capacitance value is chosen for both model and experimental validation.

2.3 Circuit Asymmetric CT-Type Topology

A 17-level asymmetrical CT-type MLI with active current paths (*Red line*) is illustrated in Fig. 3 and different switching states are tabulated in Table 1. A 17-level configuration has DC-links voltage of *R* and *L* modules 3E ($V_{C1R} = V_{C2R} = 3E$) and E ($V_{C1L} = V_{C2L} = E$), respectively.

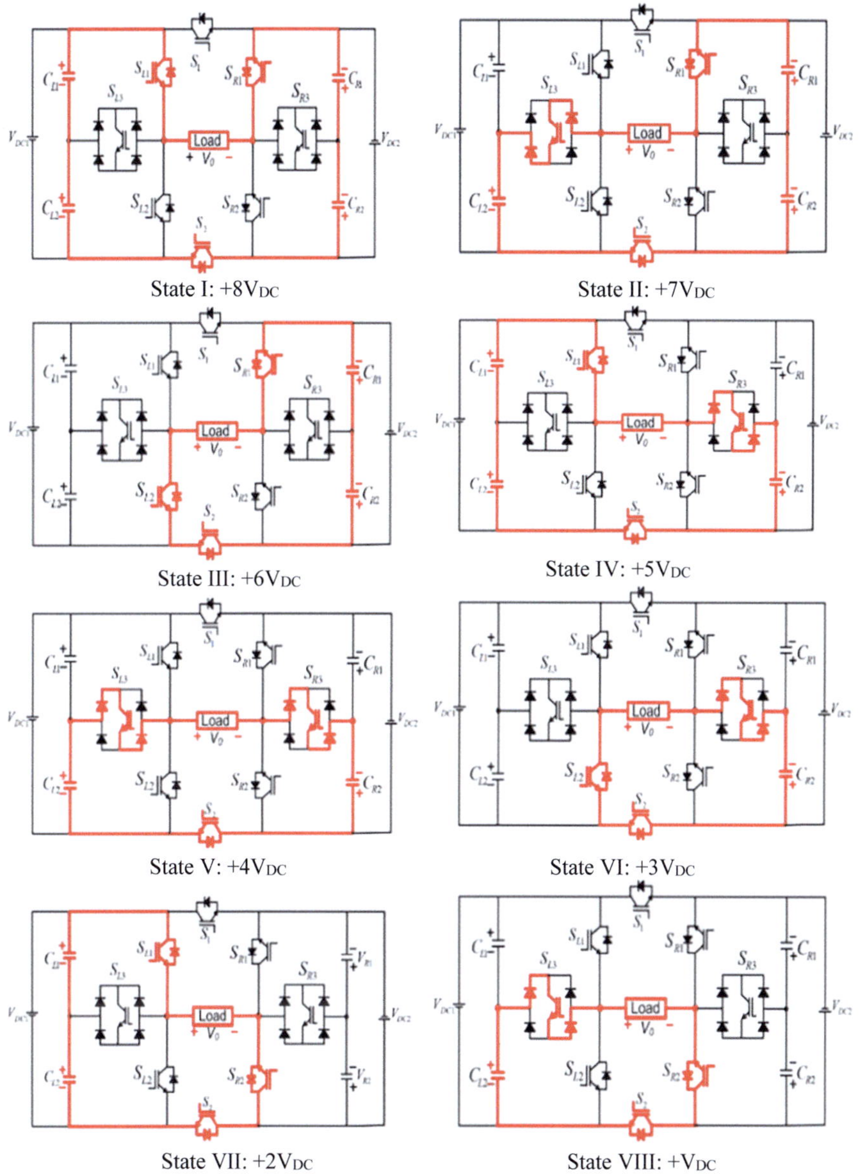

Fig. 3 Different switching states

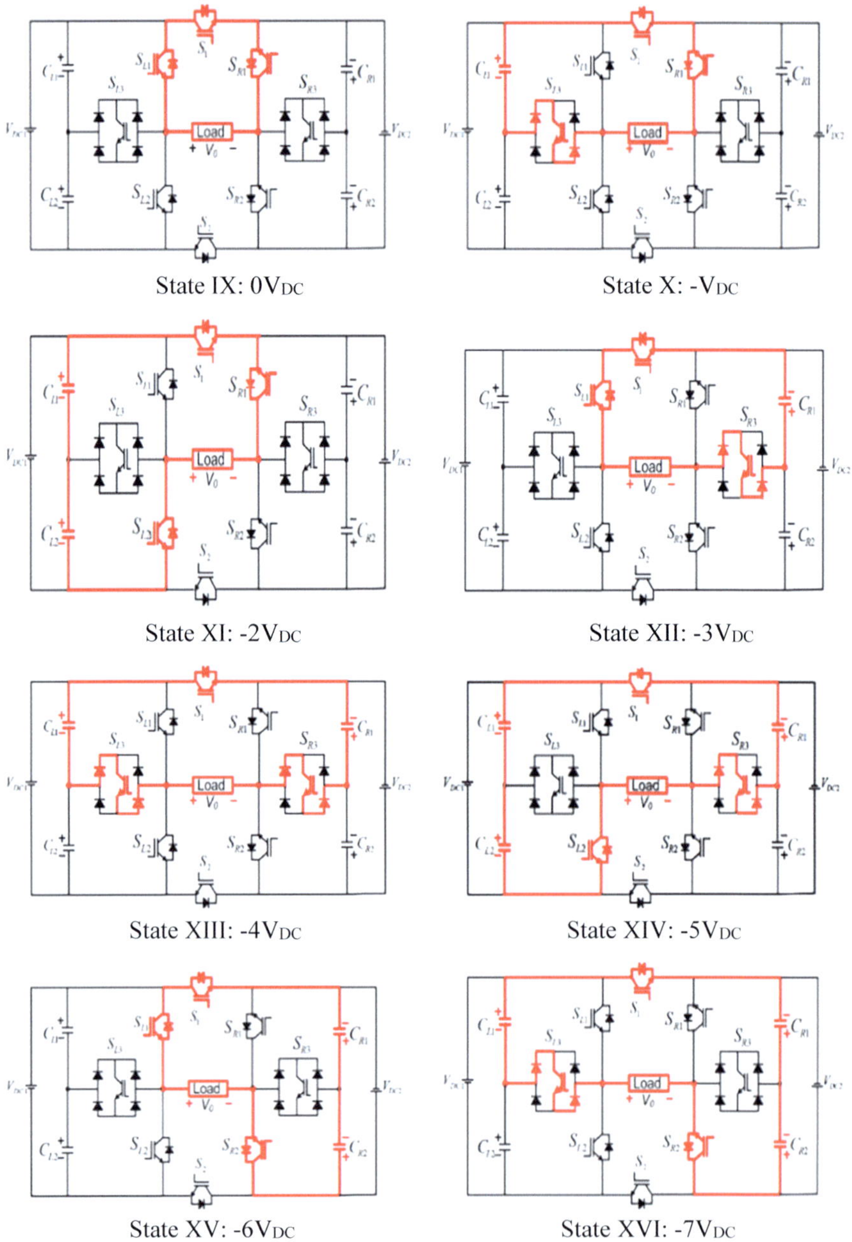

Fig. 3 (continued)

Fig. 3 (continued)

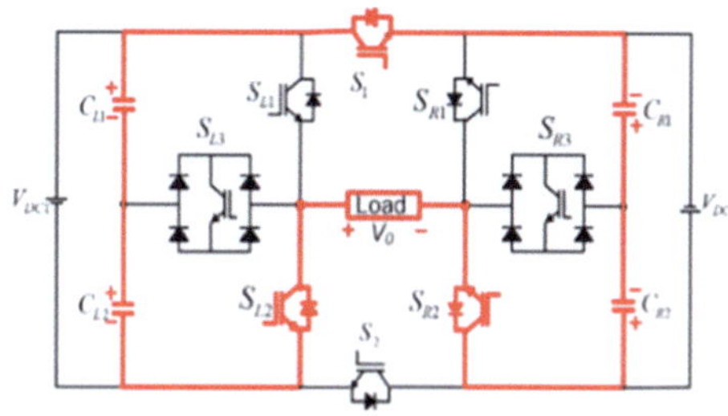

State XIV: -8V_{DC}

2.4 Voltage Stress Calculation

TSV of asymmetric operation can be estimated as the switches' maximum voltage standing capacity in the OFF state. The sum of these voltages is used to calculate the TSV of the proposed CT-Type MLI. Further, the calculation can be expressed as follows:

$$V_{S1} = E + E = 2E,\ V_{S2} = E + E = 2E,\ V_{S3} = 3E + 3E = 6E \tag{8}$$

$$V_{S4} = 3E + 3E = 6E,\ V_{S5} = E + E + 3E + 3E = 8E, \tag{9}$$

$$TSV = V_{S1} + V_{S2} + V_{S3} + V_{S4} + V_{S5} + V_{S6} + V_{S7} + V_{S8} = 36E \tag{10}$$

3 Simulation Study

3.1 Modulation and Control Method of MLI

Different modulation techniques are used for MLIs, including high switching frequency techniques (i.e., multi-carrier pulse width modulation and space-vector pulse width modulation) and low switching frequency techniques (i.e., active-harmonic elimination, selective-harmonic elimination and nearest level control). With appropriate modification, the proposed CT-type topology can be modified using any of these techniques. The level-shift pulse width modulation is adopted in this work. In the LS-PWM approach, the triangular carrier signals are contrasted with the sinusoidal reference signal. Gate pulse is generated for switching devices at various voltage levels. The peak value of the waveform is V_{ref}, peak and modulation index (MI) will be obtained using Eq. (11).

Table 1 Switching states of 17-level asymmetric configuration

States	S_{L1}	S_{L2}	S_{L3}	S_1	S_2	S_{R1}	S_{R2}	S_{R3}	O/P	C_{L1}	C_{L2}	C_{R1}	C_{R2}
1	1	0	0	0	1	1	0	0	+ 8E	Discharging	Discharging	Discharging	Discharging
2	0	0	1	0	1	1	0	0	+ 7E	–	Discharging	Discharging	Discharging
3	0	1	0	0	1	1	0	0	+ 6E	–	–	Discharging	Discharging
4	1	0	0	0	1	0	0	1	+ 5E	Discharging	Discharging	–	Discharging
5	0	0	1	0	1	0	0	1	+ 4E	–	Discharging	–	Discharging
6	0	1	0	0	1	0	0	1	+ 3E	–	–	–	Discharging
7	1	0	0	0	1	0	1	0	+ 2E	Discharging	Discharging	–	–
8	0	0	1	0	1	0	1	0	+ E	–	Discharging	–	–
9	1	0	0	1	0	0	1	0	0	–	–	–	–
10	0	0	1	1	0	1	0	0	−E	Charging	–	–	–
11	0	1	0	1	0	1	0	0	−2E	Charging	Charging	–	–
12	1	0	0	1	0	0	0	1	−3E	–	–	Charging	–
13	0	0	1	1	0	0	0	1	−4E	Charging	–	Charging	–
14	0	1	0	1	0	0	0	1	−5E	Charging	Charging	Charging	–
15	1	0	0	1	0	0	1	0	−6E	-	–	Charging	Charging
16	0	0	1	1	0	0	1	0	−7E	Charging	–	Charging	Charging
17	0	1	0	1	0	0	1	0	−8E	Charging	Charging	Charging	Charging

$$MI = \frac{V_{ref.peak}}{5V_{car}} \tag{11}$$

The proposed circuit employs the phase opposition disposition (POD) approach to create gate pulses. Figure 4a shows the schematic diagram of the control technique, and Fig. 4b shows time domain waveform of LS-PWM. Due to the switching redundancy of particular voltage levels, a specific switching state is chosen to reduce the higher to lower voltage levels switching of the IGBT. PWM signals generated for switches S_1 to S_7 are decided by the location of high-frequency carriers.

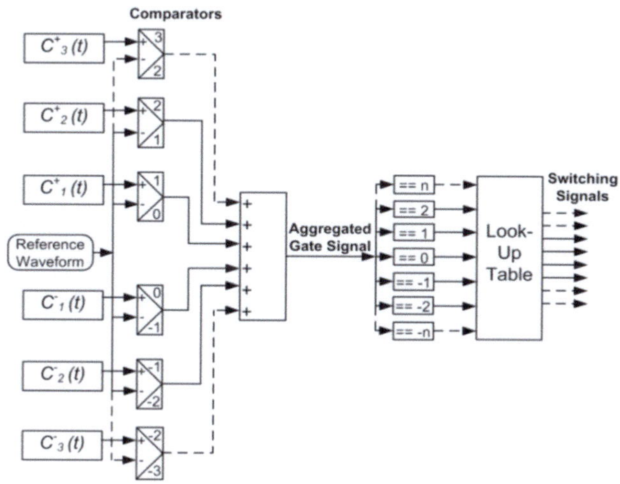

(a)

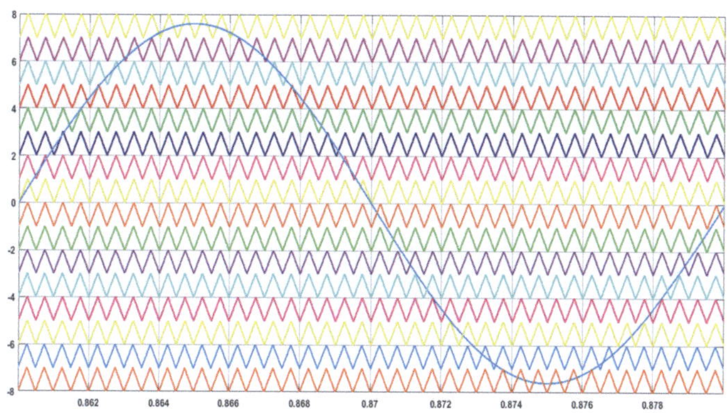

(b)

Fig. 4 **a** Block diagram of control method **b** Corresponding waveform

Table 2 Simulation parameters

Parameter	Value
Rated DC voltage	$V_{DC1} = 24\,V\,\&\,V_{DC2} = 24\,V$
Capacitors	$C_1 = C_2 = 2200\mu F$
Modulation index	$m_a = 0.85$
Load value	L = 20 mH, R = 20 Ω
Modulating wave frequency	$f_m = 50$ Hz
Switching frequency	$f_s = 3150$ Hz

3.2 *Simulation Results*

MATLAB/Simulink software is used in this sub-section to implement the proposed CT-type MLI with LSPWM modulation. The parameters of simulation are tabulated in Table 2. In the asymmetrical configuration, the input source $V_{DC1} = 24$ V and $V_{DC2} = 72$ V is applied to get the desired output. Figure 5a illustrates the 17-level voltage output waveform for the proposed topology in an asymmetric configuration. The peak value of output voltage is $V_{BB} = 96$ V at 17 levels of voltage. The proposed CT-type topology load current waveform in an asymmetric configuration is shown in Fig. 5b. Figure 6a illustrates the voltage across capacitors (C_{L1} and C_{L2}) in an asymmetric configuration. Similarly, Fig. 6b shows the voltage across capacitors (C_{R1} and C_{R2}) in an asymmetric configuration.

4 Experimental Setup Results

A laboratory prototype has been developed with power switch modules, gate drivers and isolated power supplies to validate the proposed CT-type topology. Figure 7 shows a block diagram of the experimental setup. The values of various components and parameters are enlisted in Table 3. dSPACE-1104 controller is used for implementing the simulated gate pulses and control signal on an experimental setup.

Asymmetric model is constructed by providing $V_{C1L} = V_{C2L} = 24$ V and $V_{C1R} = V_{C2R} = 72$ V. The inverter's DC-link voltage is 96 V in total. A single-phase RL load with R = 40Ω & 20Ω and L = 40mH & 20mH has connected to the module output.

The prototype is created utilizing four equal DC supplies in asymmetric operation, with $V_{C1L} = V_{C2L} = V_{C1R} = V_{C2R} = 12$ V, bringing the total DC-link voltage of the inverter to 48 V. In steady-state and dynamic load conditions, experimental results of capacitor voltage, output voltage, and load current are illustrated in Figs. 11 and 12. Figure 9a and b show 7.2% of voltage THD and 1.3% of current THD (Figure 8).

Fig. 9Experimental results with load (R = 20Ω, L = 20 mH) showing: a Output voltage THD b Output current THD

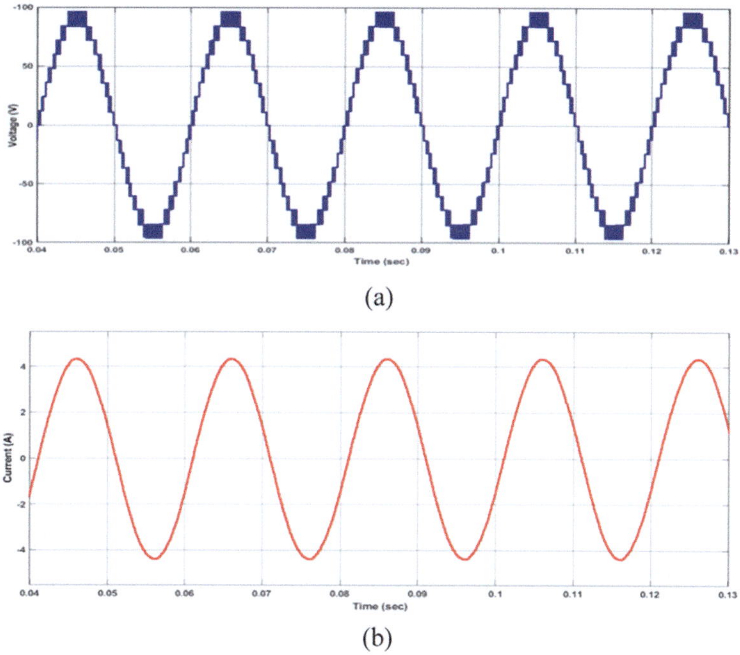

Fig. 5 **a** Output voltage **b** Load current

5 Loss Calculation

In the suggested CT-type topology, power switches have primarily been linked to two types of losses: switching and conduction losses. Conduction losses (P_C) are caused by current flow in switches while they are in the ON state. The IGBT and anti-parallel diode in the MLI switch are both traversed by the corresponding load current. As a result, P_C is determined by adding the conduction losses in the IGBT and anti-parallel diode ($P_{C,\ D}$). We estimate $P_{C,\ IGBT}$ and $P_{C,\ D}$ as follows:

$$P_{C,IGBT} = n_{IGBT}(t).\left[\frac{1}{2\prod}\int_{0}^{2\Pi}\left[V_{ON,IGBT}I + R_{ON,IGBT}I^{\beta+1}d(\omega t)\right]\right] \quad (12)$$

$$P_{C,D} = n_D(t).\left[\frac{1}{2\prod}\int_{0}^{2\Pi}[V_{ON,D} + R_{ON,D}I^2 d(\omega t)\right] \quad (13)$$

$$P_C = P_{C,IGBT} + P_{C,D} \quad (14)$$

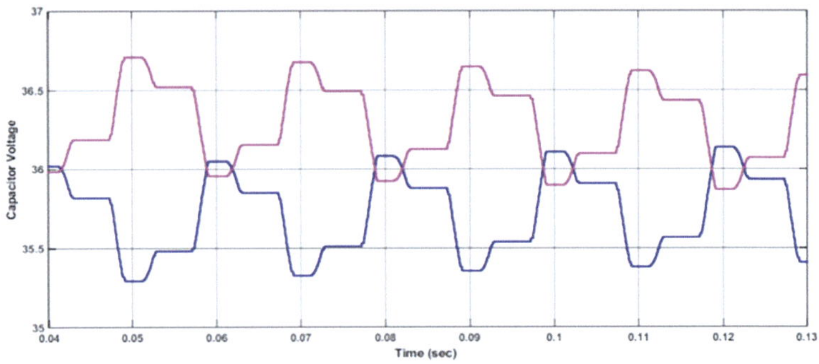

(a)

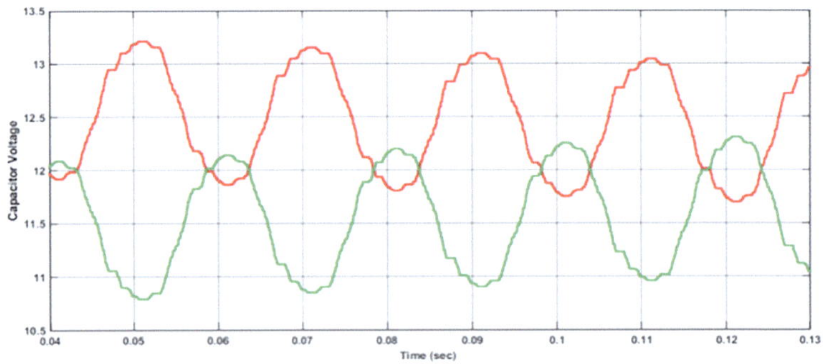

(b)

Fig. 6 Voltage across capacitors **a** C_{L1} and C_{L2} **b** C_{R1} and C_{R2}

6 Comparison Assessments

In A comparison of numerous topologies with characteristics analogous to the proposed topology has been given in this section. The proposed CT-Type MLI topology is compared with standard topologies, such as diode clamped, FC-MLIs, CHB, and topologies in [28]. It is observed that CT-Type MLI provides significantly higher output level as compared to other topologies found in the literature. It indicates that, compared to other topologies, the total number of components required to produce a given number of levels is significantly lower in the proposed topology; this can be seen in Tables 4. For example, the FC-MLI, CHB-MLI, diode clamped, and topologies components needed for the asymmetric CT-Type MLI to generate 17-level output are 22, which is substantially lower than those required in conventional topologies and topologies given in [22]. Since total standing voltage (TSV) is much higher in CT-Type MLI than in standard topologies, we must use switches with larger reverse blocking capability. Cost factor (CF) for the proposed MLI to determine the Cost-effectiveness is obtained using Eq. (15),

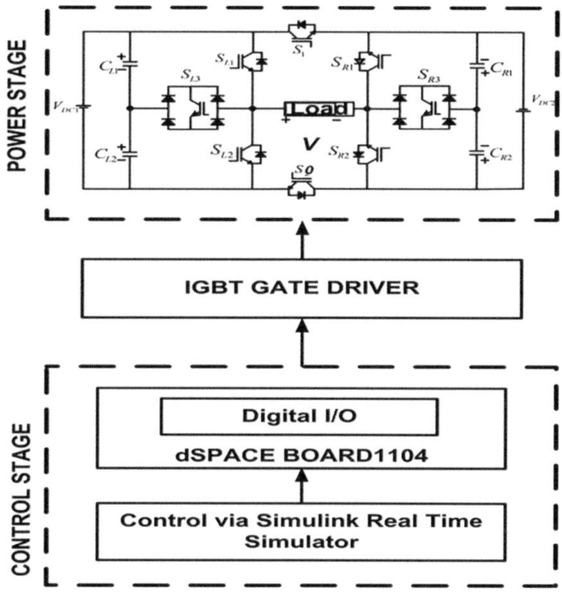

Fig. 7 Block diagram of the experimental setup

Table 3 Parameters of experimental setup

Parameter	Value
Capacitor (LGU2A222MELA)	2200 μF
DC Source (lead acid battery)	28 Ah
Driver circuit (TLP250(F))	10–35 V
Output voltage frequency	50 Hz
Switching frequency	3150 Hz
IGBT switch (GW30NC120HD)	1200 V, 30A
Digital storage oscilloscope (DL-750E)	6-channel
Controller board (dSPACE-1104)	–
Single-phase power analyzer (Fluke)	–
Load (RL Load)	R = 40 Ω, L = 40 mH R = 20 Ω, L = 20 mH

$$CF = \frac{[N_{dc} + N_s + N_{dio} + N_{cap} + \alpha\,(TSV)] * nV_{dc}}{N_{levels}} \tag{15}$$

where 'N_{dio}' is the total number of diodes, 'N_S' is the total number of switches, 'N_{cap}' is capacitor count, 'N_{DC}' is the number of DC supplies, 'α' is the weight coefficient and 'TSV' is the total standing voltage.

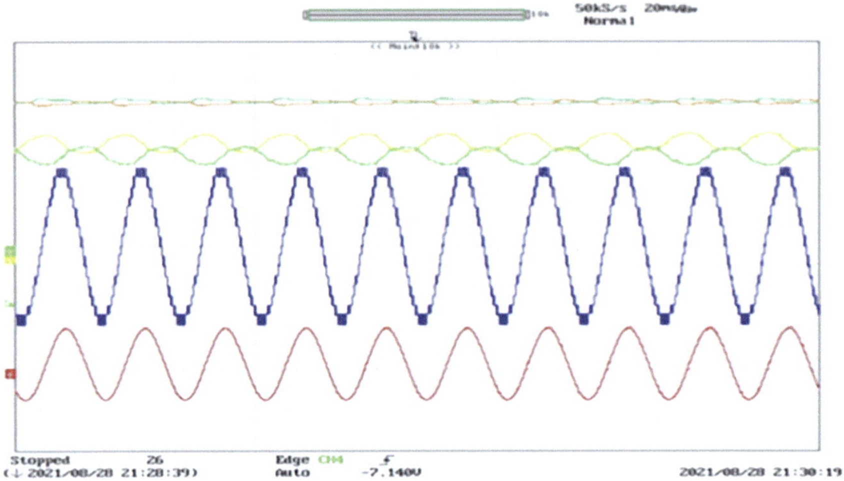

(a)

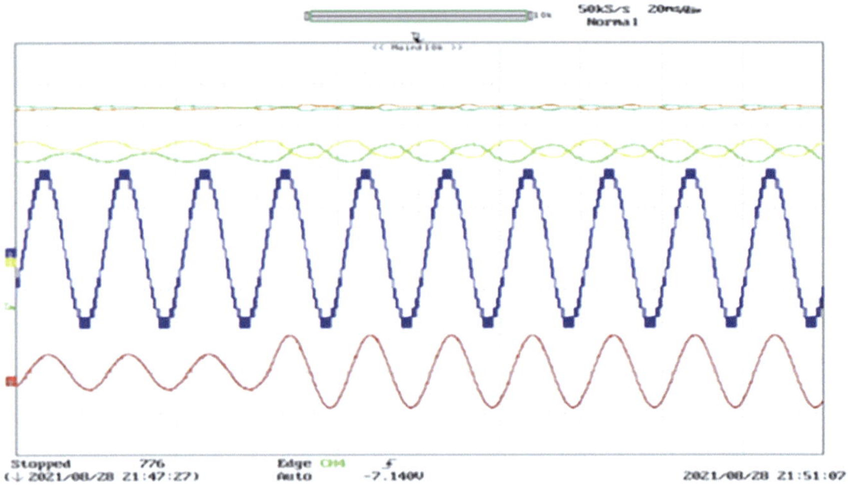

(b)

Fig. 8 Experimental results **a** capacitors voltage (C_{L1} *and* C_{L2}),, load current, and output voltage with (R = 20Ω, L = 20 mH) **b** capacitors voltage of (C_{R1} *and* C_{R2}),, load current, and output voltage with (R = 40Ω, L = 40 mH)

Fig. 9 Experimental results with load (R = 20Ω, L = 20 mH) showing: **a** Output voltage THD **b** Output current THD

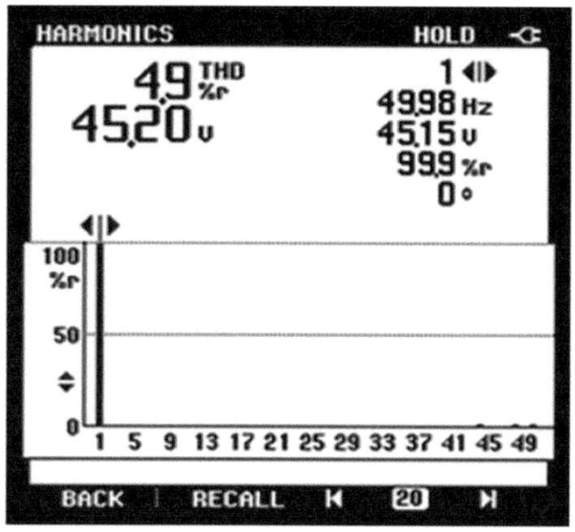

(a)

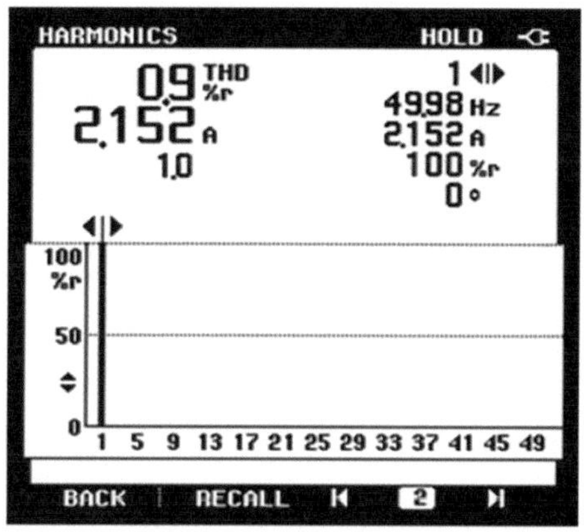

(b)

Table 4 Comparison of topologies (MLLs)to generate 17-level output based on the component requirement

Topologies		Diode clamp MLI	Cascaded H-bridge MLI	Flying capacitor MLI	[27]	[28]	Proposed CT-type MLI
Capacitors and DC supplies		16	8	16	4	4	6
Diodes		48	32	32	12	14	8
Power switches		32	32	32	12	10	8
Driver circuit		32	32	32	12	10	8
TSV		32	32	32	40	36	36
Total components		96	72	80	28	28	22
CF/ Level	= 0.5	6.57	5.18	5.63	2.82	2.7	2.47
	= 1	7.51	6.1	6.59	4	3.76	3.52
	= 1.5	8.48	7.04	7.51	5.17	4.82	4.58

7 Conclusion

This paper presents a single-phase CT-Type MLI topology with reduced device count. Also, the operating principles, control design and TSV are explained in detail. The DC-link capacitors achieve self-voltage balance. The main benefit of the proposed work is its capability to generate negative voltage levels without using a modified H-bridge. The proposed topology can be used for high output voltage waveforms without putting additional stress on the power switches. The cascading of modular structures is used to extend the proposed topology for high-voltage levels (17 levels and 289 levels) with a reduced device count. Therefore, this CT-Type MLI is suitable for renewable energy systems and high-voltage/power applications. Several simulations and experimental results of the proposed CT-Type MLI at different load conditions are presented to show the effectiveness of the topology. Several other recently developed MLI topologies are compared with the proposed topology to prove its superiority. The proposed CT-Type is superior to the other recently developed MLI topologies on the basis of number of drivers, capacitors, DC sources, and power switches, thereby making the entire system simpler and less expensive. Hence, the proposed topology's performance in various operational situations is satisfactory.

References

1. Rathore V, Yadav KB, Dhamudia S (2021) An implementation of nine-level hybrid cascade multi-level inverter (HCMLI) using IPD-topology for harmonic reduction. In: Nanoelectronics, circuits and communication systems: proceeding of NCCS 2019 (pp. 203–210). Springer Singapore
2. Gupta KK, Jain S (2014) Comprehensive review of a recently proposed multilevel inverter. IET Power Electronics 7(3):467–479
3. Kumar D, Nema RK, Gupta S (2020) A comparative review on power conversion topologies and energy storage system for electric vehicles. Int J Energy Res 44(10):7863–7885
4. Rathore V, Yadav KB (2022) Comparative efficiency analysis of five-level dual three-phase multilevel inverter fed six-phase induction motor drive. Int J Numer Model Electron Networks Devices Fields 35(3):e2981
5. Rathore V, Yadav KB (2022) Mathematical modeling and numerical analysis of SPIM drive using modified SVPWM technique. ECTI Trans Electr Eng, Electron, Commun 20(2):152–162
6. Kumar D, Nema RK, Gupta S (2021) Development of a novel fault-tolerant reduced device count T-type multilevel inverter topology. Int J Electr Power Energy Syst 132:107185
7. Kumar D, Nema RK, Gupta S (2021) Development of fault-tolerant reduced device version with switched-capacitor based multilevel inverter topologies. Int Trans Electr Energy Syst 31(7):e12893
8. Gupta KK, Ranjan A, Bhatnagar P, Sahu LK, Jain S (2015) Multilevel inverter topologies with reduced device count: A review. IEEE Trans Power Electron 31(1):135–151
9. Anand V, Singh V (2020) Compact symmetrical and asymmetrical multilevel inverter with reduced switches. Int Trans Electr Energy Syst 30(8):e12458
10. Mahato B, Majumdar S, Jana KC (2019) Single-phase Modified T-type–based multilevel inverter with reduced number of power electronic devices. Int Trans Electr Energy Syst 29(11):e12097
11. Siddique MD, Mekhilef S, Shah NM, Sandeep N, Ali JSM, Iqbal A, Ahmed M, Ghoneim SS, Al-Harthi MM, Alamri B, Salem FA (2019) A single DC source nine-level switched-capacitor boost inverter topology with reduced switch count. IEEE Access 8:5840–5851
12. Gupta KK, Jain S (2013) A novel multilevel inverter based on switched DC sources. IEEE Trans Industr Electron 61(7):3269–3278
13. Alishah RS, Hosseini SH, Babaei E, Sabahi M (2016) Optimal design of new cascaded switch-ladder multilevel inverter structure. IEEE Trans Industr Electron 64(3):2072–2080
14. Alishah RS, Nazarpour D, Hosseini SH, Sabahi M (2014) Reduction of power electronic elements in multilevel converters using a new cascade structure. IEEE Trans Industr Electron 62(1):256–269
15. Babaei E, Laali S, Bayat Z (2014) A single-phase cascaded multilevel inverter based on a new basic unit with reduced number of power switches. IEEE Trans Industr Electron 62(2):922–929
16. Gautam SP, Sahu LK, Gupta S (2016) Reduction in number of devices for symmetrical and asymmetrical multilevel inverters. IET Power Electronics 9(4):698–709
17. Siddique MD, Iqbal A, Memon MA, Mekhilef S (2020) A new configurable topology for multilevel inverter with reduced switching components. IEEE Access 8:188726–188741
18. Saeedian M, Adabi J, Hosseini SM (2017) Cascaded multilevel inverter based on symmetric–asymmetric DC sources with reduced number of components. IET Power Electronics 10(12):1468–1478
19. Salem A, Van Khang H, Robbersmyr KG, Norambuena M, Rodriguez J (2020) Voltage source multilevel inverters with reduced device count: Topological review and novel comparative factors. IEEE Trans Power Electron 36(3):2720–2747
20. Vemuganti HP, Sreenivasarao D, Ganjikunta SK, Suryawanshi HM, Abu-Rub H (2021) A survey on reduced switch count multilevel inverters. IEEE Open J Ind Electron Soc 2:80–111
21. Samadaei E, Kaviani M, Bertilsson K (2018) A 13-levels module (K-type) with two DC sources for multilevel inverters. IEEE Trans Industr Electron 66(7):5186–5196

22. Routray A, Singh RK, Mahanty R (2020) Selective harmonic elimination in hybrid cascaded multilevel inverter using modified whale optimization. Int Trans Electr Energy Syst 30(4):e12298
23. Babaei E, Farhadi Kangarlu M, Sabahi M (2014) Extended multilevel converters: an attempt to reduce the number of independent DC voltage sources in cascaded multilevel converters. IET Power Electronics, 7(1), pp157−166
24. Pulikanti SR, Agelidis VG (2011) Hybrid flying-capacitor-based active-neutral-point-clamped five-level converter operated with SHE-PWM. IEEE Trans Industr Electron 58(10):4643–4653
25. Omer P, Kumar J, Surjan BS (2020) A review on reduced switch count multilevel inverter topologies. IEEE access 8:22281–22302
26. Meraj ST, Hasan K, Masaoud A (2017) Simplified one dimensional space vector modulation of cross-switched multilevel inverter. In: 2017 IEEE 15th Student conference on research and development (SCOReD) (pp. 350–355). IEEE
27. Babaei E, Laali S, Alilu S (2014) Cascaded multilevel inverter with series connection of novel H-bridge basic units. IEEE Trans Industr Electron 61(12):6664–6671
28. Hosseinzadeh MA, Sarebanzadeh M, Rivera M, Babaei E, Wheeler P (2020) A reduced single-phase switched-diode cascaded multilevel inverter. IEEE J Emerg Sel Top Power Electron 9(3):3556–3569

Multilevel VSI for Battery Powered Applications with Novel PWM Control Scheme

Shailesh Kumar Gupta

Abstract In the industry, multilevel inverters are preferred for its application in high voltage and power. With this increase in the voltage level, the inverter's output gets better with fewer harmonic distortions. Multilevel inverters are usually used in a high-power system. As such, when there is an increase in the level of voltage, there is complexity added to the power circuit also. For resolving the challenge a unified PWM method through space vector modulation is devised for a dual 2-level 5-phase inverter system. The study is a research of all aspects of the dual 2-L 5-phase VSI, which deals with power circuit diagram using an open-end winding, the generation of vector with the appearance of the reference vector in all three regions, the time distribution of voltage vectors, the plane of voltage vector, the voltage selection for the two inverters, and the switching sequence of the two inverters in all three regions. To validate the obtained results, MATLAB/SIMULINK models based on the SVPWM scheme are developed. The output of the dual 2-level inverter system is analyzed in terms of fundamental voltage and total harmonic distortion. Finally, the output is investigated for different isolated DC sources present in both the inverters. The findings have been presented in terms of the fundamental voltage and percentage THD.

Keywords Dual 2-L 5-Phase inverter · MATLAB/SIMULINK model · Open end winding (OEW) · Space vector pulse width modulation (SVPWM) · Total harmonic distortion (THD)

1 Introduction

Among the many uses for 3-phase VSIs in ac motors are controlled rectifiers, ac motor drives, and grid-connected systems. Nonetheless, the challenges related to high-power applications can be addressed by increasing the number of phases. Compared

S. K. Gupta (✉)
Department of Electrical and Electronics Engineering, IIMT Engineering College, Meerut, India
e-mail: drskggbu@gmail.com

Y. Gopal et al. (eds.), *Multilevel Inverter Technology for Renewable Energy Systems*, Green Energy and Technology, https://doi.org/10.1007/978-3-032-03589-9_2

to their three-phase counterparts, poly-phase machines have many advantages, such as lower torque amplitude and pulsation and increased frequency, lower power in each inverter leg, smaller volume for the same output power, lower DC link current harmonics, and lower copper losses in the rotor [1].

Generally speaking, the five-phase drive system has characteristics that make it more favored and fortunate than the three-phase drive system going forward. Recent studies have also found it to be the most intriguing. For the machine drive system, the five-phase torque ripple is smaller than the three-phase torque ripple. Because the five-phase drive uses less current per leg for the same power requirements without raising the voltage of the legs, it is more appropriate for high power applications than the three-phase system and has exceptional dependability under fault modes [2–6].

However, because of advantages including better output voltage waveforms, reduced EMI emissions, and larger power ratings, multilevel inverters have grown in popularity as a substitute for conventional 2-L 3-phase VSIs in recent decades. Multilevel inverters are used to reduce the inverter's output harmonic content in high-power applications with low switching frequency. As the output level increases, the power circuit's complexity and cost increase rather than providing these advantages of multilayer topologies.

Additionally, a 3-L inverter increases the computational complexity while producing space vectors that resemble those produced by a double 2-L inverter, where each sector is divided into four distinct triangular areas depending on variables such as the angle and amplitude of the reference voltage vector [7].

Reactive power demand and harmonic impurity are undesirable phenomena in power system operation brought on by high power non-linear and time variable loads, such as rectifiers, adjustable speed drives, and office equipment like computers and printers. Variable nonlinear currents drawn, especially by high-power single-phase rectifier loads, significantly alter the outputs of nonlinear loads. The permanent magnet synchronous motor drive, which is driven by a double inverter with a DC supply, is thus used. Because potential zero sequence current causes circulating current in the winding and increases losses and current strain in power semiconductor components, it is crucial to assess it in the OEW drive system. The reference voltage is synthesized using zero sequence switching configurations in the current technique, which do not provide a sequence voltage of zero value. Zero sequence current in the OEW is eliminated by using unified SVPWM driven by nonlinear loads. Furthermore, the double SVPWM's overall switching frequency was lowered by one-third. In order to reduce total harmonic distortion, power electronic drives also use the alternate inverter PWM.

The architecture used in the 3-L single inverter system required the installation of additional capacitors and diodes, which required voltage balancing, even though the number of switches in the 2-L double inverter and 3-L single inverter systems is comparable. A double inverter system that eliminates the requirement for additional diodes or capacitors can be used to overcome this issue. The ripples are much lessened when a twin 2-L inverter system functions as a 3-L inverter. Acoustic noise and EMI emissions from the motor are also significantly decreased when a twin 2-L inverter system operates as a 3-L inverter [8, 9]. A double 2-L inverter system using SVPWM technology is proposed to address this because of its benefits, and the switching analysis of this method is investigated.

In order to cover the entire range from 85 to 100% of the maximum permitted output voltage, the authors in [10] detailed various SVPWM strategies and suggested a new method utilizing big and medium SVPWM. In [11, 12], various continuous PWM techniques for five-phase VSI have been demonstrated. Getting a sinusoidal output voltage is the goal of these approaches. The ideal switching pattern for eliminating low order harmonics in a five-phase inverter was demonstrated by the findings. In [13], the SVPWM equivalent algorithm was proposed. In this way, complicated real-time computing was avoided. In order to reduce total harmonic distortion (THD), a developed approach was proposed in [14]. Numerous studies have been conducted on multiphase machines to use twin 2-L inverter systems to create multilevel output, which has many benefits [15–21]. The DC voltage was used up to 100% in this method.

Dual 2-L 5-phase VSI and SVPWM for dual 2-L 5-phase inverter systems are covered in this article. The SVPWM approach for driving the 5ph-VSI is further examined in this article using both large and medium voltage vectors. The MATLAB/SIMULINK environment has been used to simulate the output result. The output of a dual 2-L 5-phase inverter system is investigated using THD and fundamental voltage.

2 Dual 2-L 5-Phase VSI

A double 2-L inverter setup is a great alternative, even though single inverter systems are the most widely used in many commercial processes due to their many advantages, which include improved fault tolerance, significantly reduced ripples, low current rating, higher controllability, high power density, less electromagnetic emission, and dramatically improved efficiency in applications like fishing boats and others.

The circuit diagram for a double 2-L 5-phase VSI with OEW is shown in Fig. 1. Two types of converters—a positive end converter and a negative end converter—are included in the circuit. As seen in the above image, both converters are ended at the load terminals, which are designated as a, b, c, d, and e as well as a', b', c', d', and e', respectively. As shown in Fig. 1, the inputs of the positive and negative end converters are Vdc1 and Vdc2, two distinct DC voltage sources. The random value of the DC input ratio is strictly positive. Equation (1) illustrates the relationship between the double inverter outputs V_{01} and V_{02} and the reference voltage vector V_{ref} as

$$V_{ref} = V_{01} - V_{02} \tag{1}$$

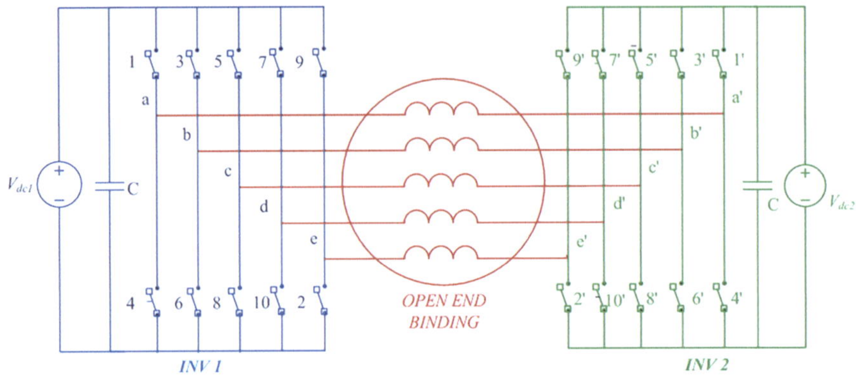

Fig. 1 Circuit diagram of dual 2-L 5-Phase VSI with OEW

3 SVPWM for Dual 2-L 5-Phase Inverter System

There are two methods for creating any circuit in standard SVPWM: one uses two big active voltage vectors, while the other uses two large and two medium active voltage vectors. Here, a compressed scheme is investigated in accordance with Eq. (1) to carry out the procedure in a double inverter system based on 2-L operation with OEW.

3.1 SVPWM Utilizing Large Voltage Vectors Only

Two large vectors, known as VP for INV1 and VN for INV2, are used to carry out the process throughout this unique scheme, with their time intervals TP and TN as

$$T_P = T_S.\frac{\sin(k\pi/5 - \alpha).|\mathrm{V}_{\mathrm{ref}}|}{\sin(\pi/5).|\underline{v}_l|} \tag{2a}$$

$$T_N = T_S.\frac{\sin(\alpha - (k-1)\pi/5).|\mathrm{V}_{\mathrm{ref}}|}{\sin(\pi/5).|\underline{v}_l|} \tag{2b}$$

$$T_0 = T_s - T_N - T_P \tag{3}$$

where, k = Number of sectors (for 5-phases, a scale of 1 to 10 is used),

$$|\underline{v}_l| = \frac{4}{5}.\cos(\pi/5).V_D = \text{Amplitude of Longer Voltage Vector Segment}$$

$\mathrm{V}_{\mathrm{ref}}$ = Reference voltage.

l = Longer voltage vector.

T_S = Sampling time and θ (0, π/5) = Angle between V_P & V_{ref}.

The sign V_{ref} represents the reference voltage using space vector, and the symbol |x| represents the modulus of a complex number x. The switching time is given by T_S, and the surrounding space vectors to the right and left of the reference space vector are represented by the indexes a and b, respectively. The indices m and l stand for medium and large space vectors, respectively. The radius of the largest circle that can be cut inside the decagon is the maximum likely fundamental peak output voltage that can be achieved using this method. The maximum fundamental peak output voltage is known as V_{max} ie $V_{max} = (2/5)2 *\cos(\pi/5)*\cos(\pi/10)*V_D = 0.61554*V_D$.Leg and phase voltages maintain their shapes when the input reference voltage varies, but their amplitude is decreased. Figure 18 demonstrates that a sizable amount of the third harmonic is present in the output phase voltage. There is also a small amount of the seventh harmonic element. The x–y components of space vectors produce these harmonic elements, which endure regardless of the reference voltage's magnitude. The x–y components are only made from the outermost decagon group of space vectors, which causes low-order distortions in the inverter's output voltages.

The only parameters that the time durations T_P, T_N, and T_0 rely on are the reference vector's amplitude and angle α. According to time periods T_P and T_N, which are limited to Ts, the double 2-L 5-phase inverter system has ten sectors in total, each with three regions in the voltage vector plane (Fig. 2). For instance, in sector I, the reference vector V_{ref} is placed in region-2 if and only if T_P is greater than Ts, as shown in Fig. 4.5b; in sector-3, if and only if T_P is less than Ts, as shown in Fig. 2c. V_{ref} is located in region-1 in any case, as Fig. 2a illustrates. If V_{ref} is not located in region-1 in accordance with the voltage second integral principle, a complementary voltage vector is needed.

Figure 2 also displays the reference voltage vector generating procedure for three distinct regions during sector 1. In this instance, two non-zero active voltage vectors are produced by both inverters. V_{1P} and V_{1N} are produced by INV1, whereas V_{2P} and V_{2N} are produced by INV2.

Following the acquisition of T_P and T_N in accordance with (2a) & (2b), time durations for each inverter are computed based on the behavior of T_P, T_N, and Ts, as shown in Table 1. It is evident that when the reference voltage vector is located

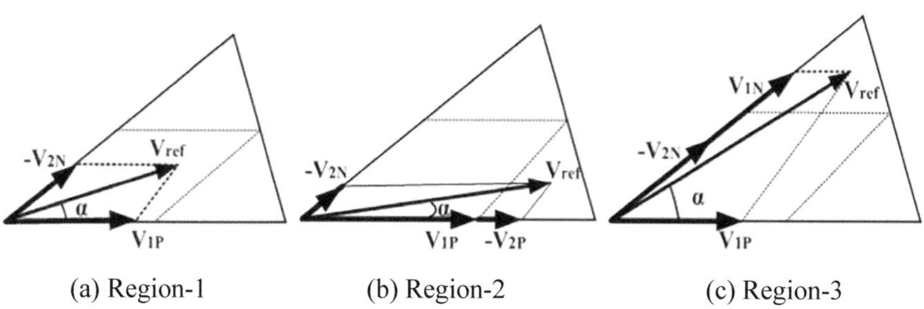

Fig. 2 Vectors synthesis procedure when V_{ref} be in all Regions

in region-1, supplementary voltage vectors are not necessary. That is, T_{1N} and T_{2P} both have zero values. Regarding region-2, time T_{1P} is bounded to Ts, and INV2 generates the analogous vector V_{2P}, which may be computed using

$$T_P \, .V_P = T_s \, . \, V_{1P} + T_{2P} \, . \, V_{2P} \tag{4}$$

$$T_{2P} = \ (T_P \, . \, V_P - T_s \, .V_{1P}) / \, V_{2P} = \ (T_P - T_s).k \tag{5}$$

$$T_N \, .V_N = T_N \, .V_{2N} + T_{1N} \, .V_{1N} \tag{6}$$

$$T_{1N} = (T_N \, .V_N - T_S \, .V_{2N}) / \, V_{1N} = (T_N - T_S)/k \tag{7}$$

The voltage vector time allocation for the double inverter in different sectors and regions is shown in Table 1. Here, the voltage vectors used at the beginning of region-1 are indicated as $V_{1P} = V_P$& $V_{2N} = V_N$. The complimentary vectors for INV1 and INV2 are V_{1N} and V_{2P}, respectively. The corresponding time durations of V_{1P}, V_{2N}, V_{1P}, and V_{2N} are denoted by T_{1P}, T_{2N}, T_{1N}, and T_{2P}, respectively.

The literature that is currently available states that switching common mode voltages are excluded if we use the voltage vectors V1, V3, V5, V7, and V9 from the positive end converter, or INV1, and V1′, V3′, V5′, V7′, and V9′ from the negative end converter, or INV2. However, we may also use the voltage vectors V2, V4, V6, V8, and V10 from the INV1 and V2′, V4′, V6′, V8′, and V10’ from the INV2 to rule out switching common mode voltages. To eliminate CMV across the load, one of these voltage vectors can be used.

Here, the voltage vectors V_{1P} and V_{2N} will be chosen from V1, V3, V5, V7, and V9 in order to carry out the plan and reduce the switching frequency. During a sample period, the appropriate compensating voltage vectors V_{1N} and V_{2P} can be chosen up from V2, V4, V6, V8, and V10. For both inverters, V0 serves as the null vector. Using the voltage vectors of both inverters that adhere to the voltage-second integral principle, Fig. 3 displays the double 2-L inverter voltage selection and vector plan for 5-phase in all ten sectors.

The inner circle (blue) and center circle (red) mark the boundaries of the several sections of a decagon that the reference vector will traverse inside a sector. Two DC supplies can be used to determine the circle's radii, which are $(V_{dc1} + V_{dc2})/\sqrt{3}$. The

Table 1 Time dispersion of voltage vectors

			INV1		INV2	
Sector I-VI			V_{1P}	V_{1N}	V2P	V2N
Region-1	$TP \le Ts$	$T_N \le T_s$	$T_{1P} = T_P$	$T_{1N} = 0$	$T_{2P} = 0$	$T_{2N} = T_N$
Region-2	$TP > Ts$	$T_N \le T_s$	$T_{1P} = T_s$	$T_{1N} = 0$	$T_{2P} = k(t_P - T_s)$	$T_{2P} = 0$
Region-3	$TP \le Ts$	$T_N > T_s$	$T_{1P} = T_P$	$T_{1N} = (T_N - T_s)/k$	$T_{2N} = T_s$	$T_{2N} = T_s$

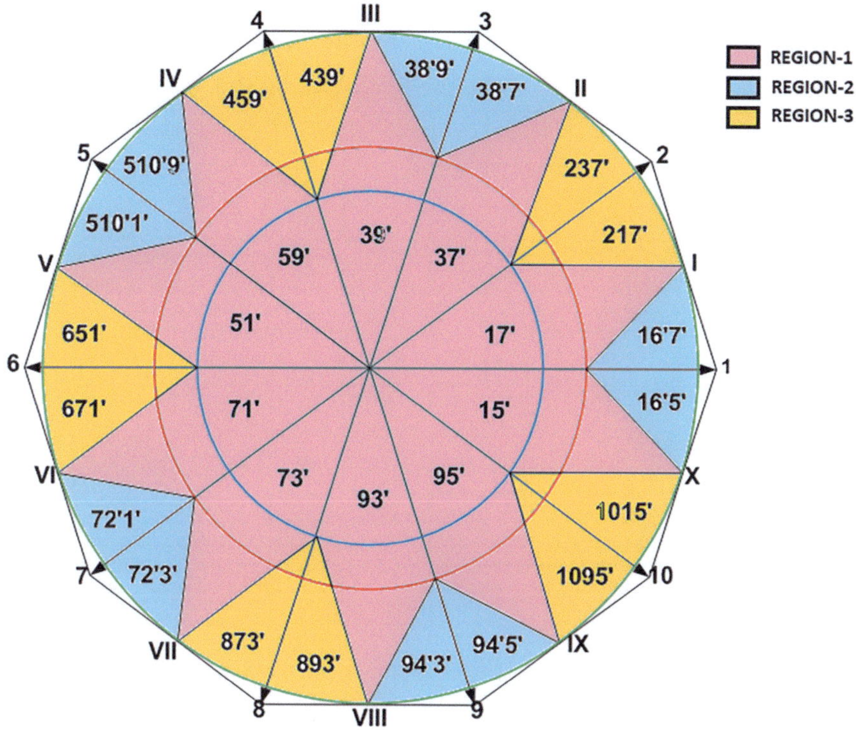

Fig. 3 Voltage selection and vector plane for dual 2-L 5-phase inverter

green circle indicates the region with the highest linear modulation index. Because the double 2-L 5-phase VSI's switching sequence lowers switching losses, switching states, and THD, it is also necessary to pay attention to it for a better understanding and successful operation of the suggested system.

Using sector-I as a sample (c), Fig. 4a, b, and c display the voltage vector switching sequences for various regions throughout the course of a sample period for both inverters. The double inverter's switching duration throughout a sample period stays at four times, regardless of where the required voltage vector is located in any given region. This period is only a third of the sampling time of a double-inverter double-SVPWM system.

The reference voltage vector will only stay in area 1 for all sectors if it is contained within the smallest circle, flipping both inverters. The voltage vector will pass through regions 1 and 3 if it is contained between the two circles. As shown in Fig. 4b, the voltage vector V1 will be clamped in area 2 of sector I, while the other two sections will be sliced.

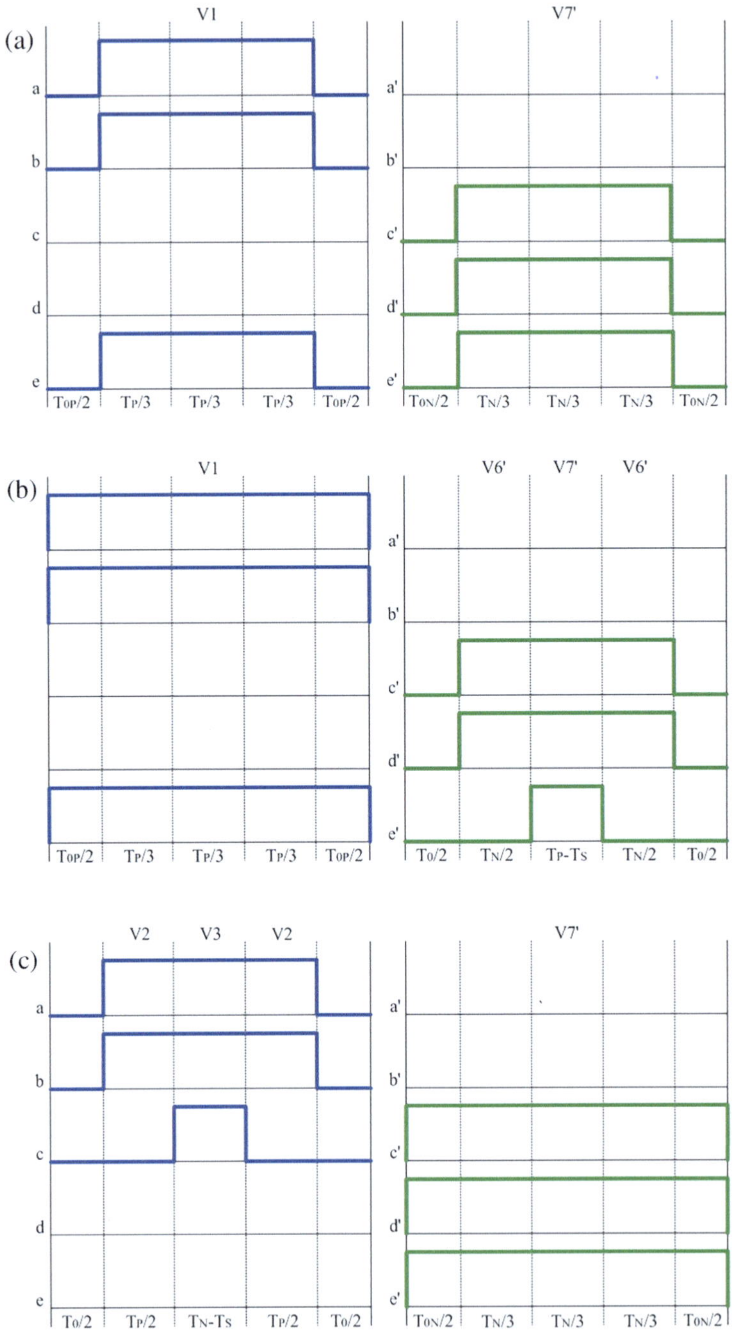

Fig. 4 **a.** Voltage vector switching sequence sector I for region-1 having scheme-1 **b.** Voltage vector switching sequence sector I for region-2 having scheme-1 **c.** Voltage vector switching sequence sector I for region-3 having scheme-1

3.2 SVPWM Utilizing Medium and Large Voltage Vectors

In the context of the phase output voltage's harmonic content, using only huge space vectors does not produce outstanding results because only two active voltage vectors are used continuously. As noted by Kelly et al. [22], a 5-phase VSI requires four active vectors instead of two. Consequently, this portion integrates the medium and large space vectors to build a new scheme. Getting sinusoidal phase voltages at the output is the goal. Conversely, the inverter can only produce pure sinusoidal voltages at its output up to a certain reference input voltage magnitude that is less than the maximum reference voltage that can be achieved with just big vectors. Consequently, the SVPWM technique explained in this section produces sinusoidal voltages at the inverter's output and is only applicable to large vectors when the reference voltage is 85.41 percent or less of the maximum fundamental. The switching pattern in Fig. 6 exhibits staggered switching in every phase, as opposed to Fig. 4, which is only appropriate for the application of large vectors. Every switch has a unique state at different times. To guarantee that every switch is turned on and off just once during each switching period, the total number of switches in each switching period stays constant.

The voltage vector plane and switching voltage vector selection for a double 2-L 5-phase double inverter are shown in Fig. 5. Scheme-01 provides a comprehensive explanation of the details. Figure 6a, b, and c one-to-one depict the necessary switching sequence for lowering the losses as well as THD in sectors I and regions 1, 2, and 3.

Only large voltage vectors can produce harmonic content in the phase voltages at the output of the inverter. Because each sector typically uses just two active vectors, we don't get adequate results. A pure sinusoidal voltage signal's total vectors must be one less than its maximum phase number. In a five-phase system, four dynamic voltage vectors must be employed rather than two in order to provide adequate sinusoidal results. Two surrounding big and two neighboring medium dynamic voltage vectors, along with one null (0) voltage vector, are used to provide reference voltage vectors for each inverter. For every given reference value and position, projections in the direction of neighboring active space vectors are the same whether just medium and large vectors are used. The duration of application for medium and large space vectors is divided proportionately (to the vector length) in order to provide sinusoidal voltage at the inverter's output.

$$T_{al} = T_a \cdot \frac{|\underline{v}_l|}{|\underline{v}_m| + |\underline{v}_l|} \tag{8}$$

$$T_{bl} = T_b \cdot \frac{|\underline{v}_l|}{|\underline{v}_m| + |\underline{v}_l|} \tag{9}$$

$$T_{am} = T_a \cdot \frac{|\underline{v}_m|}{|\underline{v}_m| + |\underline{v}_l|} \tag{10}$$

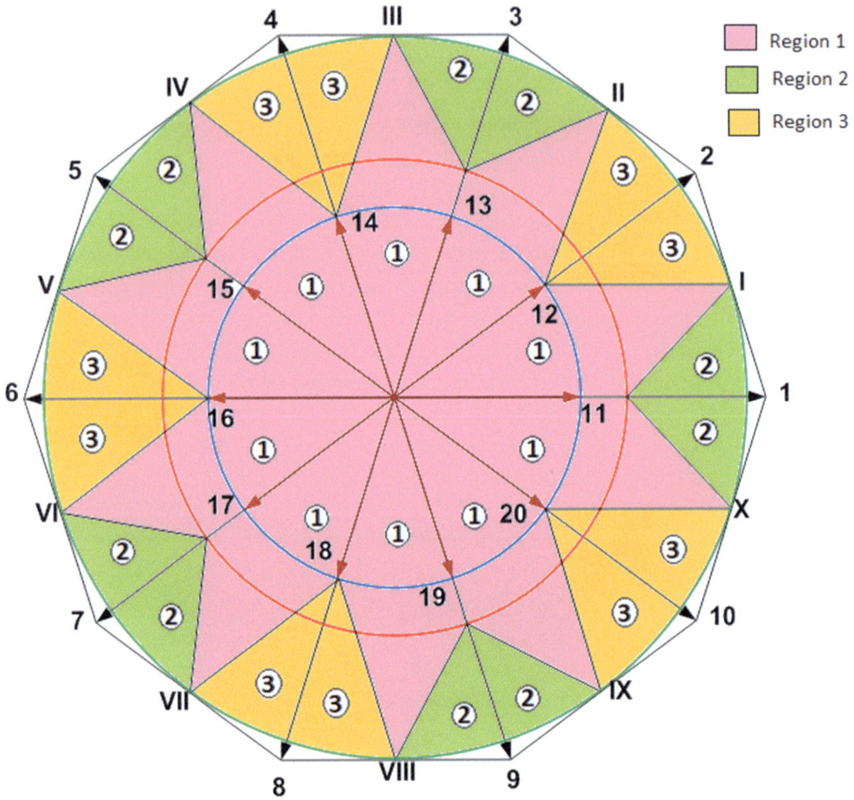

Fig. 5 Vector plane for dual 2-L 5-Phase inverter

$$T_{bm} = T_b.\frac{|\underline{v}_m|}{|\underline{v}_m| + |\underline{v}_l|} \tag{11}$$

$$T_0 = T_s - T_{al} - T_{bl} - T_{am} - T_{bm} \tag{12}$$

All the symbolic representations have their regular meaning.

4 Model for Proposed System Using MATLAB/SIMULINK

This section presents a centered model in the linear modulation limit for the step-by-step implementation of centered continuous SVPWM using MATLAB/SIMULINK [23]. Figures 8 and 9 describe the MATLAB/SIMULINK model. The block diagram representation of both methods is shown in Figs. 7, 10.

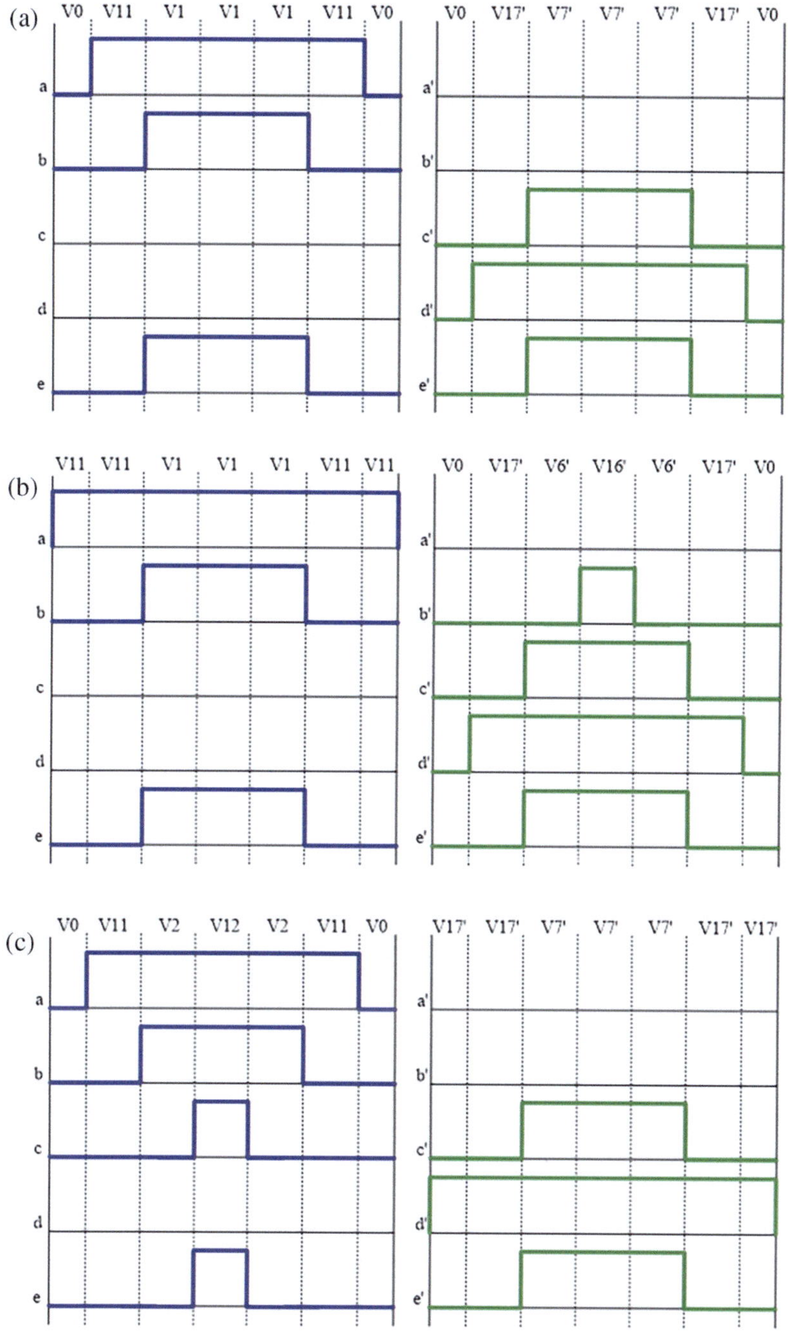

Fig. 6 **a** Voltage vector switching sequence sector I for region-1 having scheme-2 **b.** Voltage vector switching sequence sector I for region-2 having scheme-2 **c.** Voltage vector switching sequence sector I for region-3 having scheme-2

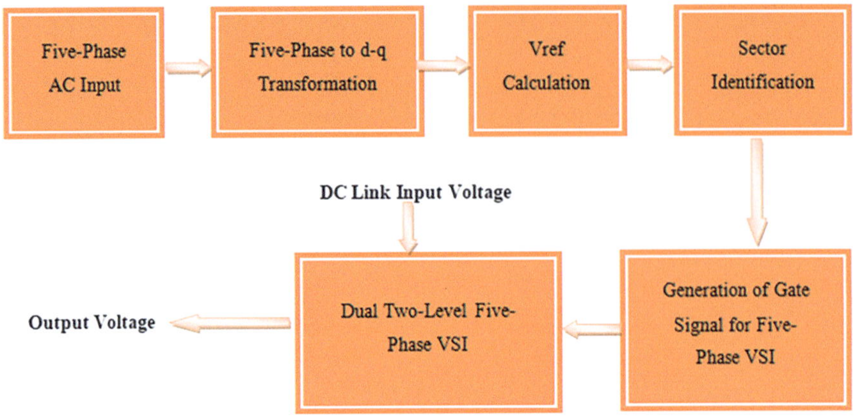

Fig. 7 SVPWM Process block diagram

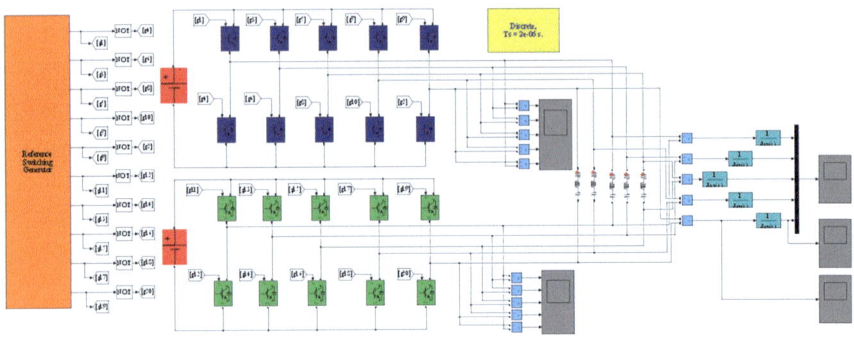

Fig. 8 MATLAB/SIMULINK Model for proposed system

The following procedures are used to create a MATLAB/SIMULINK model for a double 2-L 5-phase system that includes useful components for simulation:

1. The 'function' block from Simulink's 'Functions & Tables' sub-library is used to construct the 5-phase sinusoidal reference voltages at the inverter's input, which are 720 phase displaced from one another.
2. 'function' blocks are then used to transform the reference voltages into two-phase equivalents using Clark's conversion formulae. The 2-phase equivalent is subsequently transformed into polar form with the aid of the "Cartesian to polar" block once more from the "Simulink extras" sub-library. The magnitude and associated angle are the first and second outputs of this function block. The first output's waveform is a constant line since the magnitude value stays constant, but the second output, or angle, has a sawtooth waveform with a peak value of $\pm\,\pi$.

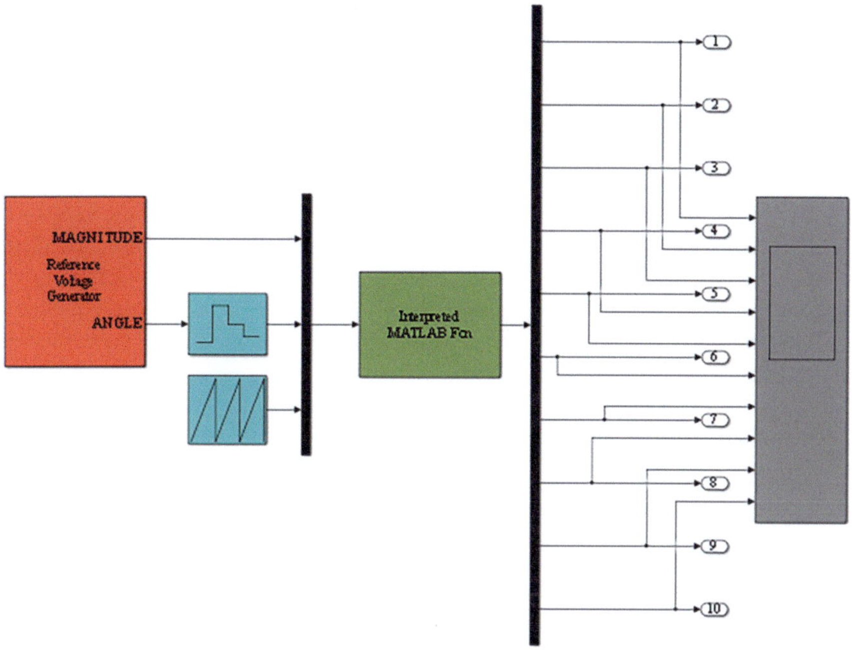

Fig. 9 Sub-Block reference switching generator

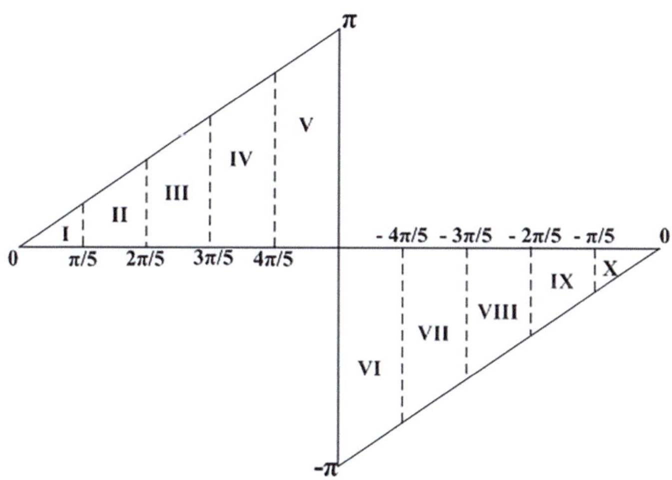

Fig. 10 Sector identification logic

3. Two-phase equivalent d-q coordinates are used to determine the sector of the reference voltage vector. The sector is identified using an angle waveform.
4. A MATLAB function block is used to determine the switching state and time for every power semiconductor switch. The timer waveform is compared to the reference's angle and magnitude using MATLAB code. The angle of the reference voltage must be fixed in order to calculate the switching time. Therefore, a block known as "zero order hold" is used to maintain a fixed angle of the reference voltage for the duration of one sampling period. This is because the width and height of the ramp time signal are the same as the inverter lag's switching time. As a result, it is created using a source sublibrary that has a "repeating sequence." Ultimately, the gating signal is acquired for every double 2-L 5-phase inverter semiconductor switch.

In the last step, all double 2-L 5-phase inverter power switches get this gating signal. Each of the five lags in this inverter has two switches. As a result, each inverter has 10 semiconductor switches in total in addition to DC input. The desired filtered line and phase voltages are set across the OEW terminal.

5 Results and Discussion

5.1 Simulation Results Using Large Voltage Vectors Only

Only two closest vectors and a null vector are used in the simulation of the double 2-L 5-phase inverter system using the newly presented PWM approach. The produced 5-phase sinusoidal input voltage signals, which are separated in phase by (2pi/5)°, are displayed in Fig. 11. Numerous studies have been conducted on multiphase machines to use twin 2-L inverter systems to create multilevel output, which has many benefits [15–21]. The DC voltage was used up to 100% in this method.

Dual 2-L 5-phase VSI and SVPWM for dual 2-L 5-phase inverter systems are covered in this article. The SVPWM approach for driving the 5ph-VSI is further examined in this article using both large and medium voltage vectors. The MATLAB/ SIMULINK environment has been used to simulate the output result. The output of a dual 2-L 5-phase inverter system is investigated using THD and fundamental voltage.

The output voltages across load from the simulation with and without a filter are shown in Fig. 12 and 13 respectively. Analysis and simulation have been performed for the different input DC voltage combinations, such as V_{dc1}:V_{dc2}. The various DC voltage combinations are displayed in Table 4.5 according to their fundamental component magnitude and THD%.

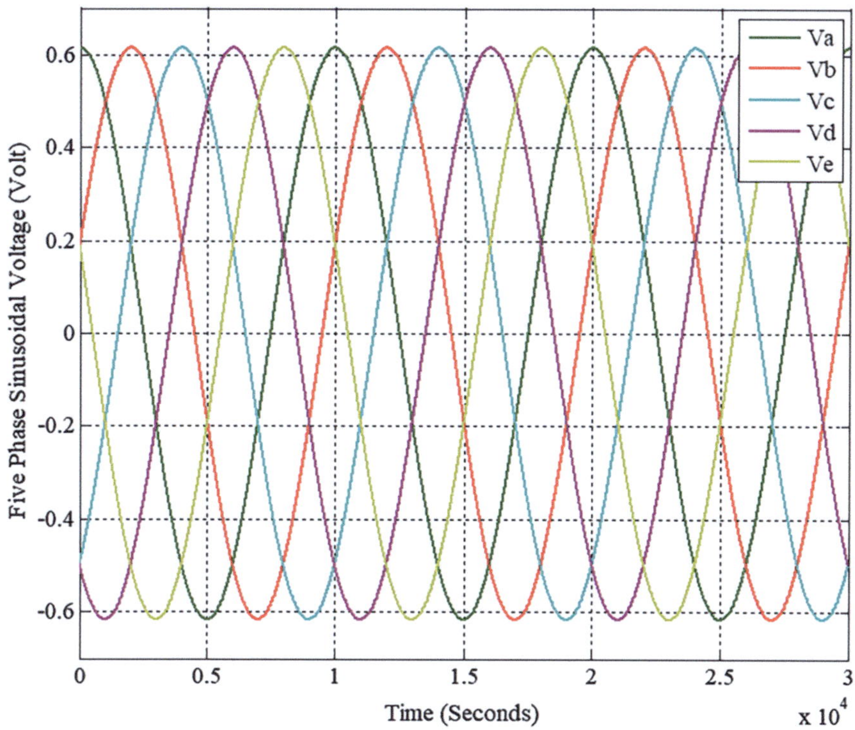

Fig. 11 Sinusoidal 5-phase input voltage signal

To provide per-unit results, the DC link voltage will be unity, with a fundamental frequency of 50 Hz and a switching frequency of 5 kHz. Using the developed model, the input reference fundamental frequency and switching frequency of the proposed inverter system are maintained at 50 Hz and 5 kHz, respectively, for continuous SVPWM. Figure 14 shows the harmonic spectrum of INV-1 for phase "a," which makes it evident that the magnitude of the fundamental is 0.5988 peak (0.4234 rms) at a frequency of 50 Hz. During the simulation, the switching frequency used was 5 kHz.

5.2 *Discussion*

According to the available literature, to generate the pure sinusoidal output, (n-1) voltage vectors are required where n represents the number of phase. Here only two large vectors are utilized instead of six due to which output is not pure sinusoidal and THD is also high.

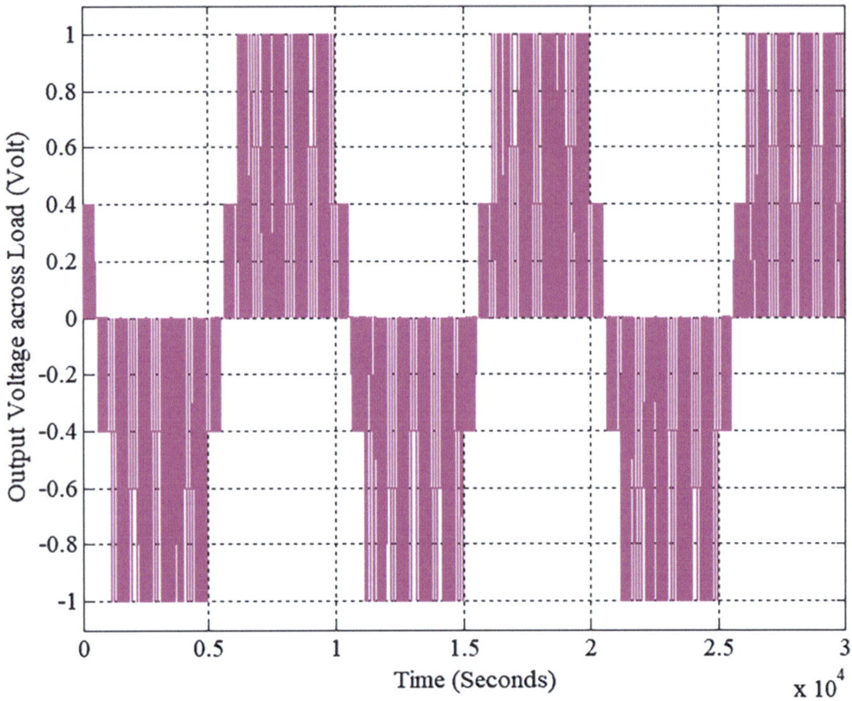

Fig. 12 Output voltages at load without filter

Table 2 displays that the % THD value is least when the ratio of V_{dc1} & V_{dc2} is 1 and equal to 24.02%. It is also clear that as both the DC link voltage increases with ratio of V_{dc1}&V_{dc2} is equal to one, magnitude of fundamental increases with same THD. When the difference of both the DC-link voltages start increasing, THD also increases but as magnitude of DC-link voltage increases with same difference, THD also starts to decrease.

5.2.1 Simulation Results Using Medium and Large Voltage Vectors

The output voltages across the load from the simulation with and without a filter are shown in Figs. 15 and 16, respectively. For the different input DC voltage combinations, such as V_{dc1}:V_{dc2}, simulation study has been performed. The various DC voltage combinations are displayed in Table 3 according to the fundamental component magnitude and THD%.

Set the fundamental frequency to 50 cycles per second, the switching frequency to 5 k cycles per second, and the DC-link voltage to unity to obtain the results in units. The suggested inverter system's input reference fundamental frequency and switching frequency are maintained at 50 cycles per second and 5 k cycles per second,

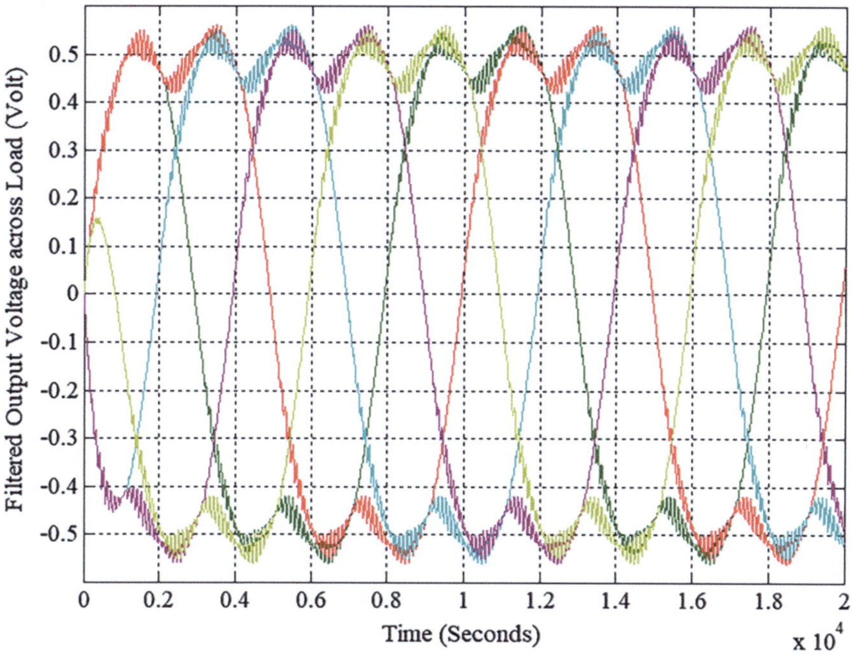

Fig. 13 Filtered output voltage across load

respectively, for continuous SVPWM during simulation using the established model. Figure 17 shows the INV-1 harmonic spectrum for phase "a." It is evident that at a frequency of 50 Hz, the fundamental magnitude is 0.5896 peak (0.4169 rms). The switching frequency used in the simulation was 5 kHz.

5.2.2 Discussion

Table 3 shows that when the ratio of V_{dc1} to V_{dc2} is 1 the value of THD is equal to 3.22%, the %THD value is at its lowest. It is also evident that the amplitude of the fundamental grows with the same THD as the DC-link voltage increases with the ratio of V_{dc1}& V_{dc2} when it equals one. THD increases when the DC-link voltage difference starts to widen, while THD decreases as the DC-link voltage difference increases in size.

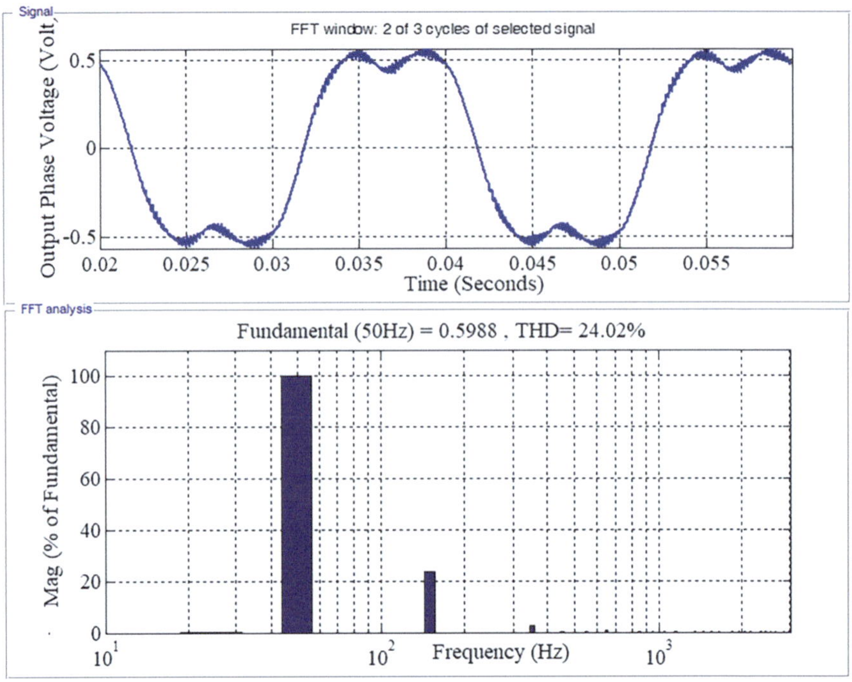

Fig. 14 Harmonic spectrum of load Voltage for particular phase 'a'

6 Outcomes

The research presented explores and presents a unified SVPWM strategy for 2-L double 5-phase inverter systems. Every sector is divided into three regions while examining the five phases. Table 4 shows how both inverters are operating in various places. The two inverters in region 1 are found to be in a switching state. Since the reference vector is in area 2, INV1 functions in a clamping mode, whereas INV2 switches. When the reference vector is in area 3, INV1 functions in a switching mode and INV2 functions in a clamping mode. Losses are kept to a minimum since one of the two inverters operates in clamping mode throughout the operation.

Table 2 Basic component & THD for various voltage combinations using scheme-1

S. no.	Voltage of INV1 (Volt)	Voltage of INV2 (Volt)	Magnitude of Fundamental Component (in P.U.)	THD (%)
1	1	1	0.5988	24.02
2	2	1	0.8983	30.15
3	3	1	1.198	36.39
4	1	2	0.8983	30.15
5	1	3	1.198	36.39
6	1	4	1.497	40.65
7	2	2	1.198	24.02
8	2	3	1.497	26.39
9	3	2	1.497	26.39
10	2	5	2.096	33.55
11	5	2	2.096	33.55
12	2	7	2.695	38.72
13	7	2	2.695	38.72
14	3	3	1.797	24.02
15	3	4	2.096	25.25
16	4	3	2.096	25.25
17	3	5	2.395	27.63
18	5	3	2.395	27.63
19	4	5	2.695	24.77
20	5	4	2.695	24.77
21	5	6	3.294	24.53
22	6	5	3.294	24.53
23	6	7	3.892	24.38
24	7	6	3.892	24.38

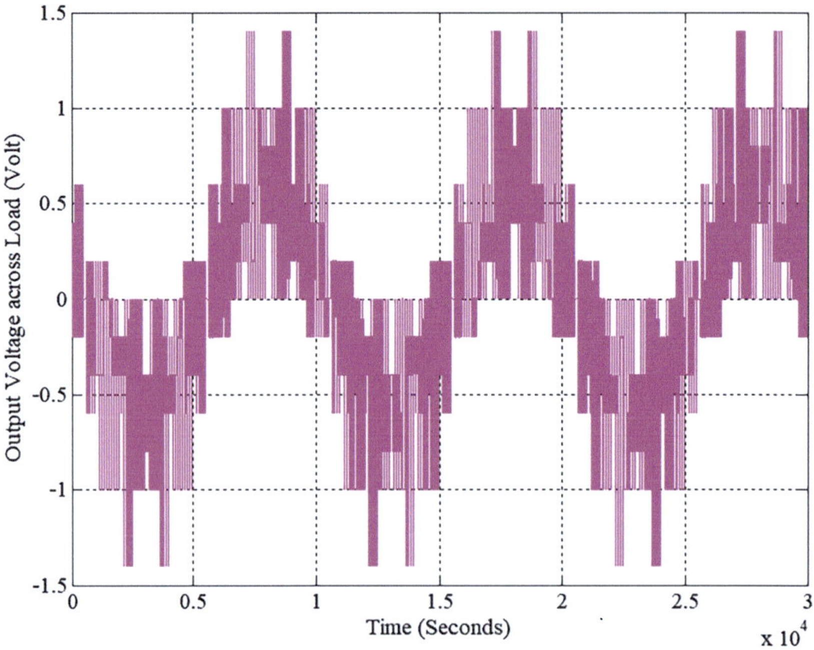

Fig. 15 Output voltages at load without filter

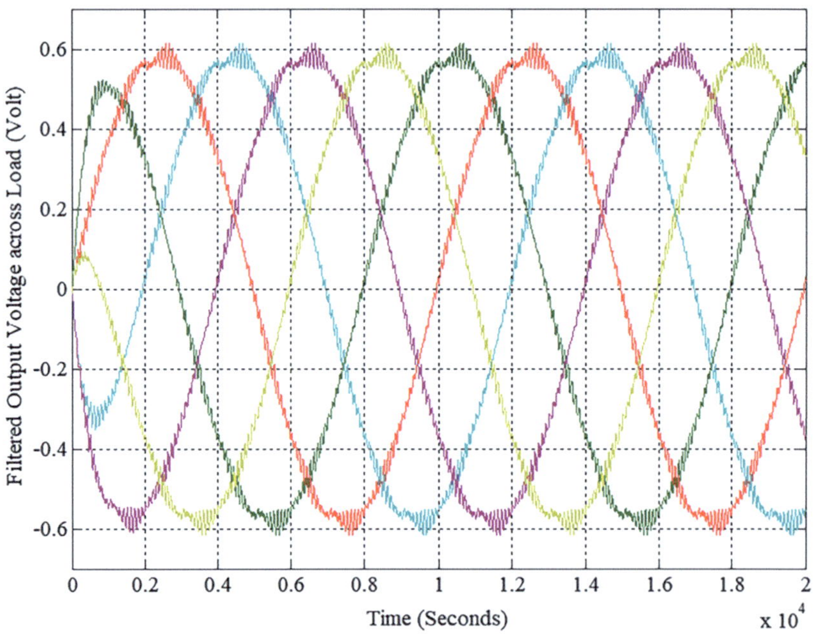

Fig. 16 Filtered output voltage at load

Table 3 Fundamental Component & THD for different Voltage Combinations with Scheme-2

S. no.	Voltage of INV1 (Volt)	Voltage of INV2 (Volt)	Magnitude of Fundamental Component (in P.U.)	THD (%)
1	1	1	0.5896	3.22
2	2	1	0.8844	15.32
3	3	1	1.179	22.70
4	1	2	0.8844	15.32
5	1	3	1.179	22.70
6	1	4	1.474	27.16
7	2	2	1.179	3.22
8	2	3	1.474	9.55
9	3	2	1.474	9.55
10	2	5	2.064	19.53
11	5	2	2.064	19.53
12	2	7	2.653	25.17
13	7	2	2.653	25.17
14	3	3	1.769	3.22
15	3	4	2.064	7.18
16	4	3	2.064	7.18
17	3	5	2.358	11.69
18	5	3	2.358	11.69
19	4	5	2.653	5.94
20	5	4	2.653	5.94
21	5	6	3.243	5.2
22	6	5	3.243	5.2
23	6	7	3.832	4.72
24	7	6	3.832	4.72

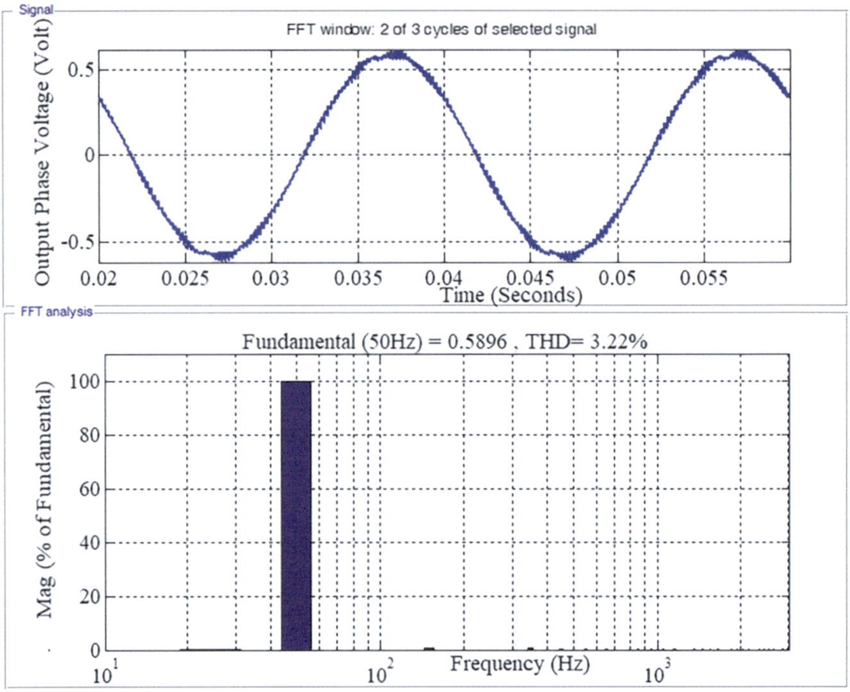

Fig. 17 Harmonic Spectrum of load voltage for particular phase 'a'

Table 4 Operation of dual 2-L 5-phase system in different states

States	Operating mode of INV1	Operating mode of INV2
1	Switching mode	Switching mode
2	Clamping mode	Switching mode
3	Switching mode	Clamping mode

7 Conclusion

This article examines two SVPWM techniques for a double 2-L 5-phase VSI in order to provide multilevel output. The first approach just generates the final output using large voltage vectors. The results clearly show that it has a large number of low-order harmonics. The results are therefore not sinusoidal. Another technique creates sinusoidal outputs by utilizing both medium and large voltage vectors. As seen in Table 5, the second method has lower levels of low-order undesired harmonics as compared to the first strategy. Both SVPWM techniques are reliable and easy to utilize.

Therefore, it can be said that for a combination of DC-link voltages where the difference between the two voltages is equal to one, the fundamental component's magnitude grows while the THD% drops as the DC-link voltage increases. As with

Table 5 Fundamental component & THD for various phases of Dual 2-L inverter system for $V_{dc1}: V_{dc2} = 1:1$

Phase description	Magnitude of fundamental component (in P.U.)	THD (%)
Five(Large Vector Only)	0.5988	24.02
Five (Large and Medium Vector)	0.5896	3.22

multilevel inverters, the power limit issue is resolved when a double 2-L inverter is used. In battery-powered applications, half of the DC voltage is produced by dividing the battery bank equally into two separate halves. As a result, in certain applications, such as electric automobiles, reliability improves. According to the investigation, when the double inverter's DC-link voltage is the same, the least THD is obtained. In the near future, experimental verification may be carried out.

References

1. Levi E, Bojoi R, Profumo F, Toliyat HA, Williamson S (2007) Multiphase induction motor drives—A technology status review. IET Electr Power Appl, 1(4), pp 489–516
2. Levi E (2008) Multiphase electric machines for variable-speed applications. IEEE Trans Industr Electron 55(5):1893–1909
3. Singh GK (2002) Multi-phase induction machine drive research—a survey. Electr Power Syst Res 61(2):139–147
4. Renukadevi G, Rajambal K (2011) Generalized model of multi-phase induction motor drive using matlab/Simulink. IEEE PES Innov Smart Grid Technol—India
5. Williamson S, Smith S (2003) Pulsating torque and losses in multiphase induction machines. IEEE Trans Ind Appl 39(4):986–993
6. Yatindra Gopal, Kaibalya Prasad Panda, Pranda Prasanta Gupta, Dheeraj Kumar Dhaked, Yogendra Arya (2023) Development and performance assessment of switched-capacitor multi-level inverters for solar PV applications. Sustain Energy Technol Assess, 79, 104360, ISSN 2213–1388
7. Chen M, Sun D (2017) A unified space vector pulse width modulation for dual two-level inverter system. IEEE Trans Power Electron 32(2):889–893
8. Yatindra Gopal, Akanksha Kumari, Kaibalya Prasad Panda, Dheeraj Kumar Dhaked, Yogendra Arya, Implementation and analysis of switched-capacitor multilevel inverters in solar photovoltaic system. Sustain Energy Technol Assess, 75, 104227, ISSN 2213–1388, https://doi.org/10.1016/j.seta.2025.104227
9. Loncarski J, Leijon M, Srndovic M, Rossi C, Gabriele G (2015) Comparison of outputcurrent ripple in single and dual three-phase inverters for electric vehicle motor drives. Energies 8(5):3832–3848
10. Iqbal A, Levi E (2005) Space vector modulation schemes for a five-phase voltage source inverter. In: 11th European conference on power electronics and applications—dresden
11. Dujić D, Jones M, Levi E (2007) Continuous carrier-based versus space vector PWM for five-phase VSI. In: The international conference on "computer as a tool"—Warsaw, Poland
12. Iqbal A, Moinuddin S (2009) Comprehensive Relationship between Carrier-Based PWM and Space Vector PWM in a Five-Phase VSI. IEEE Trans Power Electron 24(10):2379–2390
13. Gao H, Su J, Yang G, Liu J (2012) SVPWM equivalent algorithm based on carrier for five-phase voltage source inverter. In: 7th international power electronics and motion control conference (IPEMC 2012)—Harbin, China

14. Rangari SC, Suryawanshi HM, Shah B (2017) Implementation of large and medium vectors for SVPWM technique in five phase voltage source inverter. In: International conference on intelligent computing and control systems (ICICCS), IEEE
15. Gupta SK, Singh O, Khan MA, Chauhan DK (2021) Pulse width modulation technique for multilevel operation of five-phase dual voltage source inverters. J Eur Syst Autom (JESA) 54(2):371–379
16. Akanksha Kumari, Yatindra Gopal, Dheeraj Kumar Dhaked, Kaibalya Prasad Panda, Vijaya Kumar, YN (2023) A single source five-level switched-capacitor based multilevel inverter with reduced device count. E-Prime-Adv. Electr. Eng., Electron. Energy, 5, 100235, ISSN 2772–6711
17. Gupta SK (2022) THD analysis of twin two-level 5-phase inverter system for multi-level performance using two switching schemes. In: IEEE delhi section international conference on electrical, electronics and computer engineering (DELCON-2022), NSUT , New Delhi , 11–13
18. Gupta SK, Singh O, Khan MA, Chauhan DK (2021) Simulation study of five-phase dual two-level inverter system using unified space vector PWM. In: 10th IEEE international conference on communication systems and network technologies (CSNT 2021), oriental institute of science and technology, Bhopal, India , 18–19
19. Gupta SK, Singh O (2024) "Analysis of dual two-level converters for multilevel performance, multilevel converters", advances and applications (MCAA2023). Wiley-Scrivener, Scrivener Publisher, Pub
20. Gupta SK, Singh O (2025) Simulation study of time equivalent SVPWM approach for nine phase inverter system at load. In: IEEE international conference on cognitive computing in engineering, communications, sciences and biomedical health informatics (IC3ECSBHI 2025), school of engineering, gautam buddha university, Gr. Noida, India, 16–18
21. Gupta SK (2024) Switching study of twin two-level seven phase voltage source inverter system for multi-level outcome. In: IEEE—International conference on power electronics, intelligent control & energy systems (IEEE-ICPEICES-2024), Delhi technological university, New Delh, 26–28
22. Kelly JW, Strangas EG, Miller JM (2001) Multi-phase inverter analysis. In: IEEE international electric machines and drives conference, Cambridge, MA, USA pp 147–155, 17–20
23. Locment F, Semail E, Kestelyn X, Boucayorl A (2006) Control of seven-phase axial flux machine designed for fault operations. In: 32nd annual conference on IEEE industrial electronics, Paris, France, pp 1101 -1107, 6–10

Evolution of Reduced-Component Switched-Capacitor Boost Inverters for Photovoltaic System

Sumant Kumar Dalai, Kaibalya Prasad Panda, and Gayadhar Panda

Abstract The increasing global demand for renewable energy sources has accelerated the adoption of photovoltaic (PV) systems as a sustainable alternative to conventional power generation. As PV panels inherently generate low direct current (DC) voltage, inverters play a critical role in converting this DC power into alternating current (AC) suitable for grid integration. Multilevel inverters (MLIs) have emerged as a promising solution due to their ability to enhance power quality, improve efficiency, and minimize total harmonic distortion (THD). Traditional transformer-based MLIs, while effective, suffer from drawbacks such as increased size, weight, and cost. In contrast, transformer-less inverter topologies have gained significant attention for their ability to reduce system complexity, enhance power density, and improve overall efficiency. However, a major challenge in transformer-less inverters is the presence of leakage currents caused by variations in common-mode voltage (CMV) across the parasitic capacitance between the PV panel and the ground. To address these challenges, various innovative topologies have been proposed, including switched-capacitor (SC)-based inverters and hybrid configurations that optimize voltage-boosting capabilities while minimizing leakage currents. These advancements ensure improved reliability, better power quality, and enhanced grid compliance.

Keywords Multilevel inverters · Common-mode voltage · Switched capacitor · Transformer-based MLIs · Transformer-less MLIs

---

S. K. Dalai (✉)
Centre for UG & PG Studies, Biju Patnaik University of Technology, Rourkela, India
e-mail: sumantdalai.nitm@gmail.com

K. P. Panda
School of Energy Technology, Pandit Deendayal Energy University, Gujarat, India

G. Panda
National Institute of Technical Teachers Training and Research Kolkata, Kolkata, India

Y. Gopal et al. (eds.), *Multilevel Inverter Technology for Renewable Energy Systems*, Green Energy and Technology, https://doi.org/10.1007/978-3-032-03589-9_3

1 Introduction

The increasing demand for renewable energy sources has driven significant advancements in photovoltaic (PV) technology, making solar energy one of the most viable alternatives to fossil fuel-based power generation. As global energy consumption continues to rise, governments and organizations are prioritizing sustainable energy solutions to mitigate climate change and reduce carbon emissions [1]. Among renewable energy technologies, PV systems have gained prominence due to their modularity, scalability, and decreasing cost of implementation. However, one of the critical challenges in PV power generation is the efficient conversion of the inherently low direct current (DC) voltage generated by solar panels into alternating current (AC) suitable for grid connection [2].

Multilevel inverters (MLIs) play a crucial role in PV systems by improving power quality, reducing switching losses, and minimizing total harmonic distortion (THD). Compared to conventional two-level inverters, MLIs provide multiple voltage levels, resulting in smoother AC waveforms and lower electromagnetic interference (EMI). The primary classifications of MLIs include transformer-based and transformer-less topologies, each with its own advantages and drawbacks [3, 4].

Transformer-based MLIs have been traditionally employed in PV applications due to their ability to provide galvanic isolation between the PV array and the grid. This isolation enhances safety, protects against ground fault currents, and facilitates voltage matching. Transformer-based inverters can be categorized into low-frequency transformers (LFTs) and high-frequency transformers (HFTs). While transformer-based inverters provide inherent safety and compliance with grid regulations, their limitations in efficiency, size, and cost have led to a growing interest in transformer-less inverters as a more compact and energy-efficient alternative [4].

Transformer-less inverters eliminate the bulky transformer, resulting in higher efficiency, reduced cost, and smaller system size. However, removing the transformer introduces challenges related to leakage currents and common-mode voltage (CMV) variations. In the absence of galvanic isolation, CMV fluctuations across the parasitic capacitance between the PV panel and the ground can cause unwanted leakage currents, increasing safety risks, and reducing system efficiency [5].

To mitigate leakage currents, various transformer-less topologies have been proposed, including.

H5 Topology [6] introduces an additional switch to control CMV and minimize leakage current. HERIC (highly efficient and reliable inverter concept) topology uses additional bypass switches to maintain CMV stability. Neutral point clamped (NPC) and active NPC (ANPC) topologies [7] use clamping diodes or active switches to regulate voltage levels and suppress leakage currents. Switched-capacitor (SC) inverters utilize capacitors for voltage-boosting and common-mode stabilization [8]. While these solutions have improved efficiency and leakage current suppression, challenges remain in optimizing system performance, reducing component count, and enhancing cost-effectiveness.

Transformer-less inverters suffer from leakage currents due to CMV fluctuations, which can lead to safety hazards, increased electromagnetic interference (EMI), and reduced system efficiency. The elimination of transformers in grid-connected inverters introduces difficulties in maintaining a stable CMV, requiring complex control algorithms and additional switching components [9]. While transformer-less topologies offer higher efficiency than transformer-based designs, they often require additional power switches, diodes, or capacitors to manage CMV and leakage currents. This increases circuit complexity, leading to higher costs, increased control challenges, and potential reliability concerns [10].

PV inverters, particularly for residential and commercial applications, must be compact, lightweight, and capable of high power density. While transformer-less designs help achieve this, they often face thermal management issues, requiring innovative cooling mechanisms and efficient semiconductor materials [11]. Grid-connected PV systems must comply with international regulations, such as IEEE 1547, VDE 0126–1-1, and IEC 61,727, which impose strict requirements on leakage currents, THD, and voltage stability. Many existing transformer-less inverter designs struggle to consistently meet these standards without additional filtering components, which may introduce further complexity and cost [12].

Current solutions, such as common-grounded configurations, switched-capacitor networks, and active CMV control, have improved leakage current mitigation but remain limited by switching losses, additional control requirements, and increased circuit complexity. There is a need for more effective, low-cost methods to suppress leakage currents while maintaining high efficiency. The generalized block diagram layout of the proposed system is shown in Fig. 1. The CGMLI interfacing PV system with the associated control scheme has been developed with key objectives such as maximum power extraction, DC-link voltage balancing under a dynamic change in insolation, injection of clean sinusoidal grid current at unity PF, and under any grid-side perturbations (Fig. 2).

1.1 Literature of Related Work

This section summarizes the literature review on transformer-based and transformer-less MLIs. Figure 3 presents a classification of Multilevel Inverters (MLIs) based on their structural design. MLIs are broadly categorized into transformer-based and transformer-less types. Transformer-based MLIs include isolation transformers, which inherently eliminate leakage current but tend to be bulky and less efficient. On the other hand, transformer-less MLIs are more compact and efficient but require advanced topologies to address leakage current issues. These transformer-less MLIs are further divided into Neutral Point Clamped (NPC) and Common-Ground Multi-level Inverters (CGMLI). Both NPC and CGMLI can be designed to produce five-level, seven-level, or nine-level output voltages, offering improved power quality and reduced harmonic distortion.

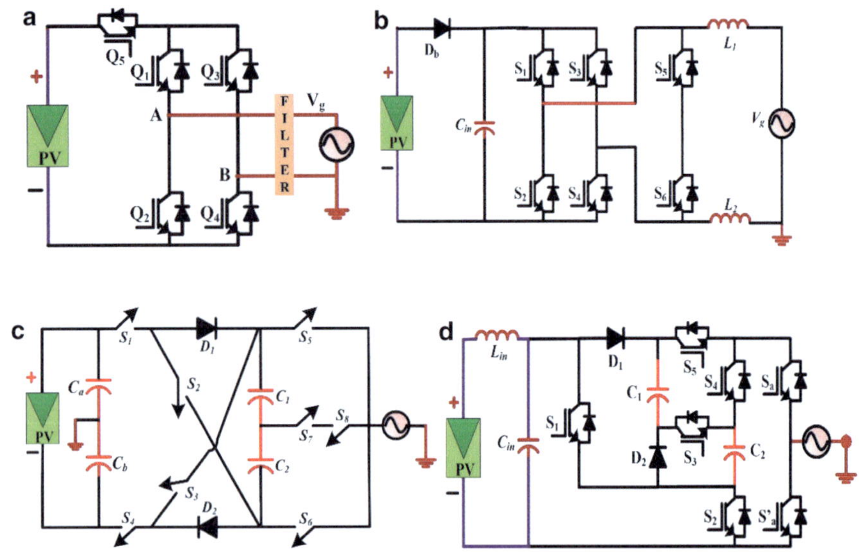

Fig. 1 Various transformer-less topologies **a** H5 [6], **b** HERIC [13], **c** NPCMLI [7], **d** SCMLI [8]

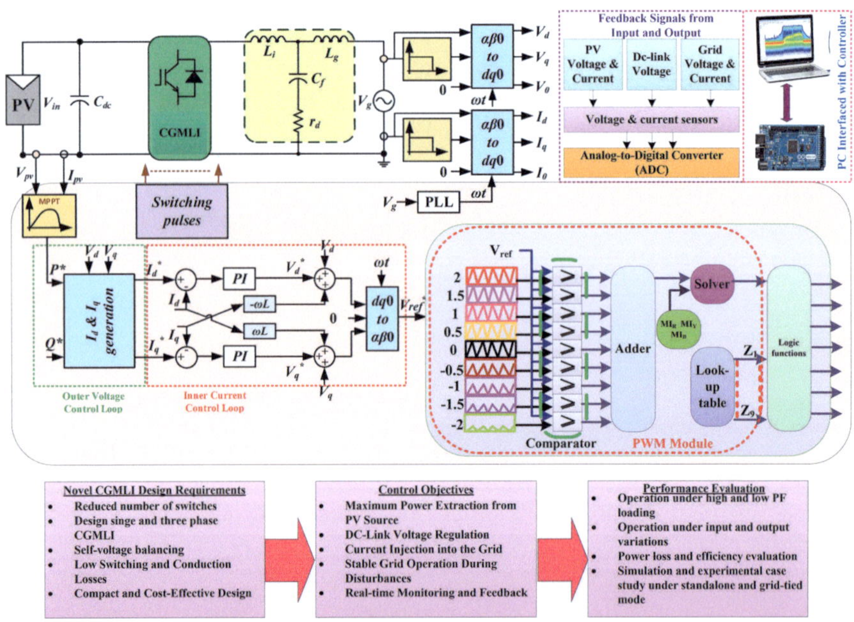

Fig. 2 Generalized block diagram of the proposed system

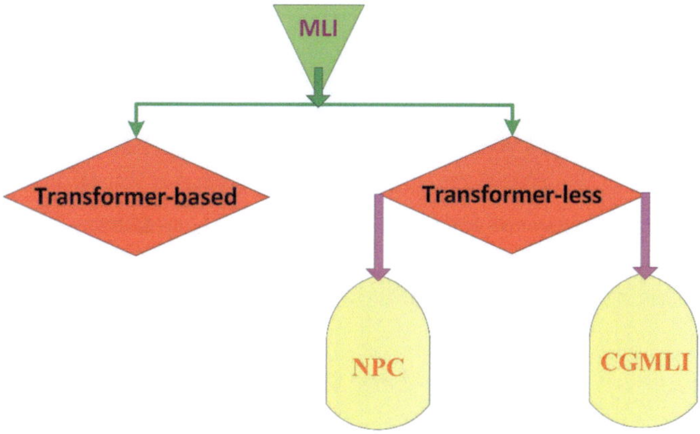

Fig. 3 Broad classification of MLI topologies

1.2 Review of Transformer-Based MLI Topologies

Multilevel inverters (MLIs) have become a crucial technology in power electronics, particularly for photovoltaic (PV) and renewable energy applications. Among various MLI topologies, transformer-based MLIs have been widely used due to their ability to provide galvanic isolation, voltage matching, and enhanced safety features. Transformer-based MLIs can be broadly classified into low-frequency transformer (LFT)-based inverters and high-frequency transformer (HFT)-based inverters. LFT-based MLIs operate at standard grid frequencies (50/60 Hz) and are commonly employed in traditional PV systems. They provide galvanic isolation, reducing the risk of leakage currents and ensuring compliance with grid safety regulations. However, they suffer from large size, heavy weight, and high core losses, making them less attractive for modern PV applications that require compact and efficient designs. Common LFT-based MLI topologies can be broadly classified into Diode-Clamped Multilevel Inverter (DCMLI), Cascaded H-Bridge (CHB) Inverter, and Flying Capacitor MLI (FCMLI).

Traditional MLIs significantly reduce system efficiency by increasing the size and cost of the converter by using extra boost stages to raise the input voltage to the necessary level. A number of SC-based MLIs have been explored and suggested for use in PV systems in recent years. In conclusion, the generation of leakage current in transformer-less grid-tied inverters, particularly in PV applications, presents a multifaceted challenge that affects system efficiency, safety, and electromagnetic compatibility. The primary source of leakage current arises from CMV fluctuations due to parasitic capacitances between the PV array and ground. As highlighted, leakage current not only diminishes the overall efficiency of the PV system but also poses significant safety risks, including the potential for electric shock and increased EMI.

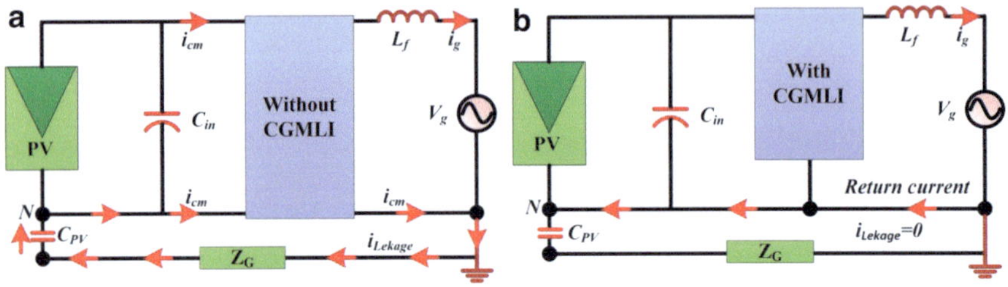

Fig. 4 Schematic of grid-tired inverter without (**a**) and with common-grounded feature (**b**)

1.3 Review of Transformer-Less MLI Topologies

Reviewing the grounded feature of transformer-less grid-tied inverters is as simple as connecting the grid's neutral point to the DC source's negative terminal [14]. The leakage current path is shown in Fig. 4, which is a typical grounded transformer-less inverter.

The Common-Mode Voltage (CMV) is not necessary in common-grounded systems because the negative terminal of the PV panels is connected to the neutral point of the AC grid [15]. To solve this problem, the authors of [16] suggested a virtual DC-link that uses five active switches. During the positive half-cycle, the input source charges the auxiliary capacitor, which is acting as the virtual DC-link. The inverter's voltage then produces the negative half-cycle. Due to the PV panel's relatively low voltage level, an additional boosting stage is required. Nevertheless, there is no guarantee that the highest voltage output will match the voltage on the grid. In place of transformers, two bipolar pulse width modulation (PWM) techniques have been suggested for use in common-grounded multilevel designs [17]. The additional inductors allow these structures to function as buck-boost. There has been a recent influx of MLIs that improve performance by using Switched-Capacitor (SC) circuits. Switched-capacitor-based grid-tied inverters offer several advantages than the inductor-based ones with a common-grounded strategy:

- **Reduced Size and Weight**: Switched capacitors are generally smaller and lighter than inductors, making the inverters more compact and easier to install.
- **Improved Reliability**: Switched-capacitor-based inverters can achieve higher efficiency due to reduced losses compared to the inductor-based designs.
- **Better Power Quality**: Switched-capacitor-based inverters can provide improved power quality by mitigating harmonic distortion and achieving better voltage regulation.
- **Faster Response**: Capacitors can respond more quickly to load and grid changes compared to inductors, resulting in faster dynamic performance of the inverter.

Overall, switched-capacitor-based grid-tied inverters offer a more efficient, reliable, and compact solution for grid-tied power conversion applications. The discussion of various strategies to limit or eliminate leakage currents underscores the need

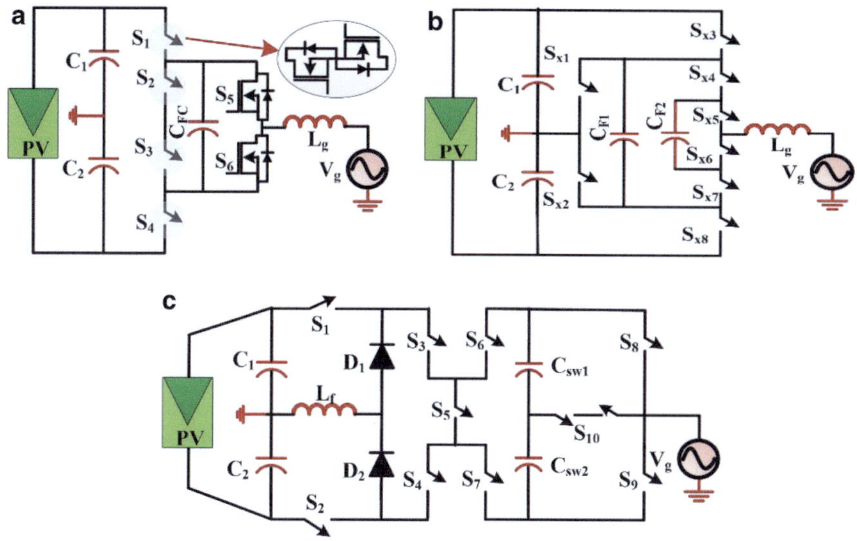

Fig. 5 Transformer-less ANPC MLIs topologies: **a** 5L-ANPC in [10], **b** 7L-ANPC in [19], **c** 9L-ANPC in [18]

for thoughtful design choices in inverter topology and control mechanisms. Methods such as common grounding, active CMV clamping, and virtual grounding have been evaluated, each.

By reducing leakage currents while keeping the system unaffected, it offers significant benefits. Enhanced electromagnetic interference (EMI) is a consequence of CMV, which raises leakage current levels. Keeping the CMV consistent, especially at the zero-voltage threshold, is critical for mitigating these difficulties. Topologies based on NPCs and Active Neutral Point Clamped (ANPCs) are one effective approach [10, 18, 19] (Fig. 5).

The low losses, reduced stress on devices, bidirectional operation, and leakage current attenuation capability of active neutral point-clamped (ANPC) multilevel converters with integrated flying capacitor (FC) technology make them widely employed in many industrial applications. Active boost neutral point-clamped (ABNPC) converters using a switched-capacitor (SC) approach have recently addressed their main limitation in the inability to utilize the full DC-link voltage at the AC output by enabling voltage boosting. However, when the supplied DC-link voltage exceeds twice the grid peak voltage, this boosting feature can result in excessive voltage stress on the semiconductor devices.

To overcome this critical issue, a dual-mode (DM) converter has been proposed, capable of maintaining high AC output quality while minimizing voltage stress across a broad range of DC input voltages. In alignment with this concept, three DM-ANPC five-level (5L) converters are introduced in [10], designed to operate efficiently in either boost or buck mode. These converters support a wide input voltage range and provide enhanced performance, leveraging a bidirectional power flow capability along with a hybrid integration of flying capacitor and switched-capacitor topologies. Each of the proposed DM-ANPC-5L converters utilizes 10 power switches to achieve these operational benefits.

In addition, due to their ability to retain most of the advantages of the original structures, hybrid topologies that combine NPC, FC, and CHB configurations have gained increasing interest for medium-power applications, such as the ANPC [20–26]. But they still necessitate a greater DC-link voltage and a lot of active and passive parts. Due to these drawbacks, it is difficult to design a novel power circuit that outperforms classical topologies in terms of component count and DC-link voltage requirements. Taking this into account, a new type of inverter called an eight-switch 7L-boost-ANPC inverter is explored and discussed in reference [19] for a wide range of general-purpose applications, including but not limited to traction, mining, fans, pumps, rolling mills, marine appliances, and grid-connected renewable energy. This results in a 50% reduction in the DC-link voltage need compared to the traditional multilevel inverter family, with an accompanying decrease in active and passive components.

Switched capacitors (SCs) have the replacement for mid-point-clamped multilevel inverters (MLIs) because of their enhanced gain. Their primary drawback is that they are hard-charged, which restricts their use to low-to-medium power levels due to the stress caused by greater device voltages. In [18], the authors suggest a nine-level inverter based on unity gain SC as a solution to this problem. This inverter's capacitors don't need any complicated control methods or external balancing control circuitry because they're self-starting and self-balancing. On top of that, using a single DC source, typical inverter topologies can achieve 100% DC-link usage. Properly positioning an inductor in the charging path also limits the peak charging current of the SCs to a safe value. In addition, the suggested inverter has fewer switching transitions, which means it has lower switching losses. The suggested inverter uses lower-rated power devices since its device voltage stress cannot exceed 50% of the input voltage (Table 1).

With their compact filter requirements, high efficiency, and reduced total switching voltage (TSV) characteristics, multilevel inverters (MLIs) have gained widespread use across various applications, primarily due to their low total harmonic distortion (THD). As the number of output voltage levels increases, the quality of the current injected into the grid improves, since a higher number of levels results in a lower THD in the output voltage waveform. The common-grounded inverters based on multilayer SC are detailed in this section. The suggested five-level common-grounded inverter without a transformer is shown in Fig. 6a in [27]. To implement this architecture, two switched capacitors and nine switches are utilized. The switched capacitors can be programmed to charge and discharge at specific states, allowing them to achieve a range of voltage levels. There are two ways the capacitors can be charged while the inverter is running: either in series and discharged in parallel, or in parallel and charged in parallel. The input voltage and the desired output voltage magnitude determine the option. Assuming $V_{C1} = V_{C2} = Vdc/2$ in buck mode and $V_{dc} = V_{C1} = V_{C2} = V_{dc}$ in boost mode, the voltage across capacitors C_1 and C_2 is half of the

Table 1 Overview of transformer-less 5L in [10]-7L in [19]-9L in [18] NPC-type MLIs for grid-integrated PV system

Category	Technical challenges addressed	Key concerns
Voltage gain and DC-link utilization	• Low voltage gain in conventional MLIs • Incomplete utilization of the DC-link voltage • Need for high DC input voltage to achieve desired AC output	• Achieved full or enhanced DC-link utilization with optimized designs • Operated efficiently with single or reduced-voltage DC sources • Suitable for PV and battery-based systems without additional boost converters
Component count and efficiency	• High number of switches and passive components in traditional designs • Increased losses and lower system efficiency due to more components	• Reduced switch count (e.g., eight-switch 7L inverter) • Improved efficiency (97.2%) • Compact, cost-effective hardware with simpler structure
Voltage stress on devices	• High voltage stress on switches and capacitors in conventional topologies • Need for high-voltage-rated devices	• Reduced device voltage stress to Vdc/2 or less • Enables the use of lower-rated, less expensive semiconductors • Enhances reliability and thermal performance
Capacitor charging and voltage balancing	• High peak charging current in SC-based topologies • Complexity in maintaining capacitor voltage balance	• Use of inductors to suppress SC charging spikes • Self-balancing capacitor structures reduce control complexity • Redundant switching states assist in FC voltage regulation
Control complexity and modulation	• Conventional schemes require complex control algorithms • Challenges in synchronizing multilevel outputs	• Simple modulation techniques (LS-PWM, PS-PWM) • Reduced controller burden • Redundant states help in balancing and reactive power support
Grid integration and adaptability	• Inflexibility with varying DC input (e.g., battery, PV) • Lack of compatibility with dynamic sources	• Dual-mode (buck/boost) operation accommodates wide DC voltage ranges • Smooth mode transitions with minimal EMI and stress • Suitable for grid-connected renewable systems

DC-link voltage. There are five possible levels of output voltage, which are 0, $\pm$ 1, and $\pm$ 2.

On the other hand, there are a number of drawbacks to all of the existing topologies, such as increased component count, total standing voltage (TSV), number of

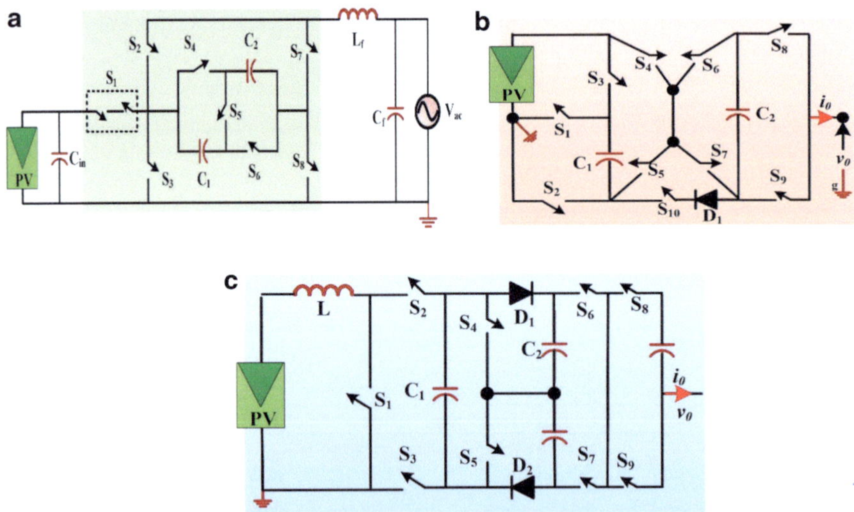

Fig. 6 Transformer-less CGMLIs topologies: **a** 5L-CGMLI in [27], **b** 7L-CGMLI in[28], **c** 9L-CGMLI in [29]

on-state conducting devices, capacitors with higher voltage ratings, and the requirement for devices with high voltage ratings. Furthermore, three or more capacitors are used by the majority of seven-level (7L) CGT topologies that have been made accessible so far [27, 30–32]. Due to its unsuitability for PV applications, the architecture [33] has limited range of non-unity power factor functioning. Here, a new 7L CGT topology with a threefold voltage gain is suggested in [28], taking into account the additional enhancements in the aforementioned topologies with a minimum capacitor count or reduced energy storage. A threefold voltage gain inverter scheme without transformers at any level is suggested here. The parasitic capacitance-induced photovoltaic ground leakage current is eliminated by the suggested inverter because of the shared ground. The suggested design improves reliability while reducing total cost by using only two capacitors with lowered voltage ratings. Less strain is exerted on the semiconductor devices by the capacitors. In addition, the capacitor voltages are self-balancing, which means they don't need any sensors to work. Reduced conduction losses are another benefit of the suggested inverter, which uses the fewest number of conducting elements possible.

The common-ground-type inverters have advanced in recent years, according to the literature [34, 35], allowing them to generate five levels of voltage at once and achieve double voltage gain. A quadruple boost topology is revealed by the 9L MLI that was introduced in [36] and [37]. Using a basic logic gate-based pulse width modulation (PWM) approach, the switching devices are tested under various load situations in [38]. Using NPC-based SC MLIs helps reduce leakage current, as mentioned in references [39]. Because of their zero common-mode voltage, which makes them ideal for solar photovoltaic applications, inverters [40] that share an alternating current (AC) and direct current (DC) ground have recently attracted a lot of attention. Present topologies have low voltage gains and discontinuous DC source currents. A new common-ground-type boost inverter is suggested in [41] that can generate continuous DC source current with a much improved dynamic voltage gain. Two power diodes, nine power MOSFETs, four capacitors, and one boost inductor make it up. In a single-stage operation, the voltage is increased and nine levels of alternating current are generated simultaneously.

Furthermore, another area of research is the design of single DC 9L-CGMLI structures that minimize stress, leakage current, and switch count. For photovoltaic (PV) applications, the 9L-CGMLIs described in [42, 29, 43] efficiently decrease leakage current. With this in mind, and in accordance with standard VDE 0126–01-01, which limits the leakage current to 300 mA, new structures have evolved with almost no leakage current at all. All of the advantages in solar PV applications may be attributed to the CG feature, which is being introduced intelligently with the SC MLIs. The CG-based construction uses a lot of switches to get a high voltage gain [44], which makes it more expensive and wastes more power. In [45], CG-based SC structures were introduced for PV applications with just 10 conducting switches. Both topologies use a bidirectional switch, however, one achieves a step-up voltage that is four times higher and the other is two times higher. More recent works [46] also reveal topologies where the CG connection uses nine switches to generate 9L output. In order to reduce inrush current and capacitor voltage ripple, the structure described in [47] is suggested for use in high-frequency AC microgrid applications. The circuit in [48] is well suited for continuous input current, and the CG-based construction in [49] can provide a high-gain output. Almost every structure that has been covered up to this point has been investigated in single-phase settings. Table 2 lists the main issues with the current CGMLIs.

2 Conclusion

By highlighting the rising global movement toward renewable energy systems, notably photovoltaic (PV) technology, as a sustainable and scalable option to satisfy expanding energy demands, this thesis was able to lay the groundwork for the research that was presented in this publication. By focusing on the requirement for high-quality power conversion, better efficiency, and minimized harmonic distortion (THD), it established the crucial role that multilevel inverters (MLIs) play in the process of interfacing low-voltage DC outputs from photovoltaic panels with the alternating current (AC) power grid. A discussion of the various transformer-based and transformer-less MLI topologies, including an analysis of their respective advantages and characteristics. There are a number of drawbacks associated with transformer-based inverters, including their bulkiness, higher cost, and lesser efficiency. However, these inverters do offer galvanic isolation and inherent safety. Transformer-less inverters, despite being more compact and efficient, effectively

Table 2 Overview of transformer-less 5L in [27], 7L in [28]-9L in [29] CGMLIs for grid-integrated PV system

Aspect	5-L [27]	7-L [28]	9-L [29]
Technical challenges addressed			
Leakage current elimination	Uses common-ground structure to eliminate leakage current	Eliminates leakage via grounded PV and grid neutral	Common AC and DC ground structure ensures zero CM voltage
Voltage boosting without extra stage	Dual-mode operation (buck/boost) achieves boosting in single stage	Achieves triple voltage gain without external boost converter	Achieves high boost ratio in single stage using dynamic inductor charging
Wide input voltage range handling	Operates efficiently between 200 and 400V input	Moderate range, focused on simplicity	Wide range with continuous input current suitable for PV
Reduced switch count and filter complexity	Nine switches and two capacitors; five-level output simplifies filter design	Minimal devices: Ten switches, one diode, two capacitors	Efficient design with nine switches, two diodes, and four capacitors
High-quality multilevel output	Five-level waveform with reduced HD (as low as 2.1%)	Seven-level output reduces harmonic distortion	Nine-level waveform improves power quality and reduces EMI
Capacitor voltage balancing	Manual design; not inherently balanced	Self-voltage balancing–no sensors needed	Natural capacitor balancing during switching
Continuous input current (good for PV)	Not guaranteed—some states disconnect PV	Not addressed explicitly	Ensured continuous current–PV friendly
Key Concerns/Limitations			
Capacitor sizing and control complexity	Requires precise control of charging discharging states	Capacitor design must ensure proper voltage ratios	Higher number of capacitors; ripple must be managed
Voltage stress on switches	Moderate (half-to-full Vdc depending on mode)	All switches 2/3 of peak voltage; low TSV	Some switches handle high voltage; overall, TSV still optimized
Switching and conduction losses	Moderate losses; addressed with thermal analysis	Reduced conduction losses via fewer active switches per cycle	Very efficient design; modeled to reach 96% efficiency
Reactive power capability	Supported in both buck and boost modes	Supports arbitrary power factor operation	Verified with both resistive and RL loads
THD and EMI mitigation	Low THD, simplified filtering	THD 3.8%; meets grid standards	Nine-level output offer slow THD, though not quantified

(continued)

Table 2 (continued)

Aspect	5-L [27]	7-L [28]	9-L [29]
Modulation complexity	Requires level-shifted PWM and logic based mode switching	Simple modulation due to balanced design	Multi-carrier PWM with dynamic boosting—more complex
Cost and volume considerations	Mode rate size and cost due to dual-mode structure	Lowest cost and volume among peers (use slow-rated components)	Compact but more capacitors and inductors may increase volume slightly

mitigate leakage current, ensuring compliance with standards such as VDE 0126–01-01. keeping this in mind, the main contribution of this thesis was the design, Neutral Point Clamped (NPC) and Common-Ground Multilevel Inverter (CGMLI) structures were presented in this chapter as a means of addressing these difficulties. These structures have gained popularity due to their capacity to reduce leakage current, achieve voltage boosting, and maintain good power quality simultaneously. Additionally, the significance of switched-capacitor (SC) designs was explored as a means of lowering the number of components, improving voltage gain, and providing support for self-balancing processes.

References

1. di Benedetto M, Lidozzi A, Solero L, Crescimbini F, Grbovi´c PJ (2018) Five-Level E-type inverter for grid-connected applications. IEEE Trans on Industry Applications, 54(5), pp 5536–5548
2. Mondol MH et al (2023) A new integrated multilevel inverter topology for renewable energy transformation. IEEE Trans Ind Appl, 59(3), pp 3031–3043
3. Chamarthi PK et al (2021) A novel three-phase transformerless cascaded multilevel inverter topology for grid-connected solar PV Appli. IEEE Trans on Indu Appl, 57(3), pp 2285–2297
4. A Ahmed 2019 An efficient single-sourced asymmetrical cascaded multilevel inverter with reduced leakage current suitable for single-stage PV systems IEEE Trans on Energy Convers 34 1 211 220
5. S Behara N Sandeep UR Yaragatti 2018 Design and implementation of transformer-based multilevel inverter topology with reduced components IEEE Trans Ind Appl 54 5 4632 4639
6. MNH Khan M Forouzesh YP Siwakoti L Li T Kerekes F Blaabjerg 2020 Transformerless inverter topologies for single-phase photovoltaic systems: a comparative review IEEE J Emerg Sel Top Power Electron 8 1 805 835
7. Guzmán et al Losses and CMV evaluation in transformerless grid-connected PV topologies. 544–548. https://doi.org/10.1109/ISIE.2009.5213296
8. Dalai SK, Panda KP, Siwakoti YP, Panda G (2024) Three-Phase neutral-point-clamped multilevel inverter interfacing grid interactive renewable energy system. In Int J Circuit Theory Appl, https://doi.org/10.1002/cta.4299
9. Dalai SK et al (2024) Three-Phase switched-capacitor boost self-balanced multilevel inverter for photovoltaic applications. In IEEE Trans Energy Convers, pp 1–10

10. Farhangi M et al (2023) A single-stage switched-boost grid-connected five-level converter with integrated active power decoupling under polluted grid voltage condition. In IEEE Open J Ind Electron Soc, 4, pp 328–345
11. R Barzegarkhoo 2023 New family of dual-mode active neutral point-clamped five-level converters IEEE Trans Power Electron 38 10 12236 12253
12. Panda et al KP (2019) Reduced switch cascaded multilevel inverter with new selective harmonic elimination control for standalone renewable energy system. IEEE Trans on Industry Applications, 55(6), pp 7561–7574
13. Albalawi et al An H5 transformerless inverter for grid connected PV systems with improved utilization factor and a simple maximum power point algorithm. Energies. 11 2912. https://doi.org/10.3390/en11112912
14. PR Kumar 2015 Seventeen-level inverter formed by cascading flying capacitor and floating capacitor H-Bridges IEEE Trans Power Electron 30 7 3471 3478
15. TDC Busarello 2018 Simplified Small-signal model for output voltage control of asymmetric cascaded H-Bridge MLI IEEE Trans on Power Electronics 33 4 3509 3519
16. Surapu PR, Sandeep N, Verma AK (2024) A Nine-Level mid-point-clamped inverter with full DC-Bus utilization and suppressed charging current for PV application. In IEEE J Emerg Sel Top Power Electron
17. YP Siwakoti 2019 A novel seven-level active neutral-point-clamped converter with reduced active switching devices and DC-Link voltage IEEE Trans on Power Electronics 34 11 10492 10508
18. Barzegarkhoo R et al (2023) A novel active neutral point-clamped five-level inverter With 1-Stage-integrated dynamic volt boosting feature. IEEE Trans on Power Elect. 38(6), pp 7796–7809
19. Gopi K et al (2022) An optimized hybrid pwm strategy for five level NPC VSI with unequal DC-Links in a PV system. IEEE J Emer Selected Topics Ind Elect, 3(3), pp 766–776
20. Abhilash T et al (2019) A seven-level VSI with a front-end cascaded three-level inverter and flying-capacitor-fed H-Bridge. IEEE Trans Ind Appl, 55(6), pp 6073–6088
21. J Liu X Zhu J Zeng 2020 A seven-level inverter with self-balancing and low-voltage stress IEEE J Emerg Sel Top Power Electron 8 1 685 696
22. MNH Khan 2020 Switched capacitor integrated (2n + 1)-Level step-up single-phase inverter IEEE Trans on Power Electron 35 8 8248 8260
23. SS Lee Y Bak S-M Kim A Joseph K-B Lee 2019 New Family of Boost Switched-Capacitor Seven-Level Inverters IEEE Trans on Power Power Electron 34 11 10471 10479
24. Lee SS, Lee K (2019) Dual-T-Type seven-level boost active neutral-point-clamped inverter. IEEE Trans Power Electron, 34(7), pp 6031–6035
25. Sathik MJ, Sandeep N, Blaabjerg F (2019) High gain active neutral point clamped seven-level self-voltage balancing inverter. IEEE Trans on Circuits and Systems II: Express Briefs, pp 1–1
26. Lee SS, Lim CS, Lee K-B (2020) Novel active-neutral-point clamped inverters with improved voltage-boosting capability. IEEE Trans on Power Electronics, 35(6), pp 5978–5986
27. T-T Tran 2020 A three-phase constant common-mode voltage inverter with triple voltage boost for transformerless photovoltaic system IEEE Access 8 166692 166702
28. P Garanayak G Panda 2016 Fast and accurate measurement of harmonic parameters employing hybrid adaptive linear neural network and filtered-xleast mean square algorithm IET Gener Transm Distrib 10 2 421 436
29. Jakhar A, Sandeep N, Verma AK (2024) Common-ground-type quadruple boosting nine-level inverter. IEEE J Emerg Sel Top Ind Electron, pp 1–8
30. Khan MNH et al. (2021) A common grounded type dual-mode five-level transformerless inverter for photovoltaic applications. IEEE Trans Ind Electron, 68(10), pp 9742–9754
31. Panda KP et al. (2020) Reduced switch count seven-level self-balanced switched-capacitor boost multilevel inverter. IEEE international conf. on power electronics, drives and energy systems, pp 1–6
32. Sahoo SK, Kumar S, Singh B (2022) Wiener variable step size with variance smoothening based adaptive neurons technique for utility integrated PV-DSTATCOM system. IEEE Trans on Ind Electron, 69(12), pp 13384–13393

33. A Parida B Subudhi 2020 Modified leaky LMS-based control strategy for reliable operation of single-stage three-phase grid-tied PV system IET Renew Power Gener 14 1453 1462
34. A Jakhar N Sandeep AK Verma 2024 Common-ground-type inverter topology with low voltage stress and boosting ability Trans Circuits Syst II: Express Briefs 71 5 2854 2858
35. Y Wang 2021 A T-Type Switched-Capacitor Multilevel Inverter With Low Voltage Stress and Self-Balancing IEEE Trans Circuits Syst I Regul Pap 68 5 2257 2270
36. Y Song B Wang 2013 Survey on reliability of power electronic systems IEEE Trans Power Electron 28 1 591 604
37. VSK Prasadarao S Peddapati B Kumar 2024 A voltage-boosting seven-level switched capacitor multilevel inverter with reduced device count IEEE J Emerg Sel Top Power Electron 12 1 743 753
38. A Khosroshahi M Abapour M Sabahi 2015 Reliability evaluation of conventional and interleaved DC–DC boost converters IEEE Trans Power Electron 30 10 5821 5828
39. A Srivastava 2023 A new nine-level highly efficient boost inverter for transformerless grid-connected PV application, IEEE J Emerg Sel Top Power Electron. 11 3 2730 2741
40. R Barzegarkhoo M Farhangi SS Lee RP Aguilera YP Siwakoti 2022 Nine-level nine-switch common-ground switched-capacitor inverter suitable for high-frequency AC-microgrid applications IEEE Trans Power Electron 37 5 6132 6143
41. MJ Sathik 2024 A nine level ANPC boost type inverter topology with reduced component stress, IEEE Trans Circuits Syst II Express Briefs 71 1 380 384
42. SS Lee CS Lim K-B Lee 2020 Novel active-neutral-point-clamped inverters with improved voltage-boosting capability IEEE Trans Power Electron 35 6 5978 5986
43. SS Lee 2022 A novel common-ground-type nine-level dynamic boost inverter, IEEE J Emerg Sel Top Power Electron. 10 4 4435 4442
44. E Babaei SS Gowgani 2014 Hybrid multilevel inverter using switched capacitor units IEEE Trans Ind Electron 61 9 4614 4621
45. PR Bana KP Panda G Panda 2020 Power quality performance evaluation of multilevel inverter with reduced switching devices and minimum standing voltage IEEE Trans Ind Inform. 16 8 5009 5022
46. A Taghvaie J Adabi M Rezanejad 2018 A self-balanced step-up multilevel inverter based on switched-capacitor structure IEEE Trans Power Electron 33 1 199 209
47. N Sandeep 2019 Switched-capacitor-based quadruple-boost nine-level inverter IEEE Trans Power Electron 34 8 7147 7150
48. S Kumari N Sandeep 2023 A self-balancing switched-capacitor-based dual boosting nine-level inverter, IEEE J Emerg Sel Top Power Electron. 11 5 5311 5320
49. Y Nakagawa H Koizumi 2019 A boost-type nine-level switched capacitor inverter IEEE Trans Power Electron 34 7 6522 6532
50. MD Siddique 2021 Single-Phase step-up switched-capacitor-based multilevel inverter topology with SHEPWM IEEE Trans Ind Appl 57 3 3107 3119

Analysis of Solar PV-Powered Reduced Component Switched-Capacitor-Based Multilevel Inverter

J. Vijaychandra, Yatindra Gopal, K. Veda Prakash, B. Vanajakshi, Lukasz Knypinski, D. Vijaya Kumar, and N. Sowjanya

Abstract Renewable energy systems have garnered significant global attention as sustainable and eco-friendly alternatives to conventional energy sources. Considering their accessibility and scalability, solar photovoltaic (PV) systems have become the most widely used technology among them. In the integration of RESs to the electric power utility grid, multilevel inverters (MLIs) play a crucial role in enhancing power quality and conversion efficiency. In this paper, a five-level solar PV-powered reduced switched-capacitor MLI is proposed and studied. In order to harvest maximum power, artificial neural network maximum power point tracking (ANN-MPPT) is used in this study and an optimal setting of ANN weights is employed using Scaled Conjugate Gradient (SCG) machine learning algorithm. The entire analysis is carried out using MATLAB/Simulink software.

Keywords Solar PV system · Switched-capacitor MLI · Artificial neural networks · Scaled conjugate gradient method · Renewable energy systems

J. Vijaychandra (✉) · K. Veda Prakash · B. Vanajakshi
Department of Electrical and Electronics Engineering, Lendi Institute of Engineering and Technology, Vizianagaram, Andhra Pradesh, India
e-mail: vijaychandrajvc@gmail.com

L. Knypinski
Faculty of Control, Robotics and Electrical Engineering, Poznan University of Technology, Poznan, Poland

D. Vijaya Kumar · N. Sowjanya
Department of Electrical and Electronics Engineering, Aditya Institute of Engineering and Technology, Srikakulam, Andhra Pradesh, India

Y. Gopal
Department of Electrical Engineering, National Institute of Technology Arunachal Pradesh, Jote, Arunachal Pradesh, India

Y. Gopal et al. (eds.), *Multilevel Inverter Technology for Renewable Energy Systems*, Green Energy and Technology, https://doi.org/10.1007/978-3-032-03589-9_4

1 Introduction

The widespread consumption of fossil fuels, which have limited resources and catastrophic impacts on the environment exemplified in global warming and climate change, has exploded exponentially as a result of population growth and modernization of technology. As a result, scientists everywhere are looking for new ways to ease back on or completely stop utilization of fossil fuels. Aesthetically, the most appealing approach is to use renewable energy since it has the potential to drastically reduce or even practically do away with the need for fossil fuels. Production as well as consumption of renewable energy has surged tremendously across the intervening 10 years [1–3].

With a million MW of renewable energy capacity deployed globally since 2012, there has been noticeable progress in technology for creating, setting up, and running new renewable projects. The International Renewable Energy Agency (IRENA) recently forecasted that 3068 GW of installed capacity will be completed by 2022. Africa accounted for 1.8% of this installed capacity, followed by Europe (21.1%), North America (14.94%), South America (8%), and Asia (47.4%), which led the most to the replacement of traditional energy sources.

By being exposed to the sun's radiant heat and light, solar energy can be collected. In addition to being plentiful—the sun emits an immense quantity of energy every day—this type of energy is renewable, which means it can be continuously regenerated. Solar energy is practically untapped and may be used as long as the sun remains illuminated, in contradiction to fossil fuels, which take millions of years to create and have a limited supply. PV technologies are moving quickly, improving the efficiency of solar cells, slicing expenses for manufacturing, and expanding their use in a variety of scenarios. These technological advances have made it possible to incorporate solar energy into our daily lives and created emerging possibilities for large-scale solar power generation.

Reduced high-order harmonics and low current ripple harmonics in induction motor drives are two additional advantages of high-frequency inverters. Considering the numerous advantages they offer, including high power density, precisely controlled heat distribution, effectiveness, and credibility, power converters are critically important for microgrids.

In consequence of their multiple structural positive characteristics, multilevel inverters (MLIs) are one of the contemporary converters and are therefore appropriate for a wide range of renewable energy applications. Due to their ability to elevate voltage and make use of a smaller amount of DC sources than conventional MLIs, switching capacitor MLIs (SC-MLIs) are the most used type of MLI. Cascaded MLIs are looked upon in order to provide greater diversity of voltage levels while aiming for fewer devices; nevertheless, this approach results in a high blocking voltage and fails to warranty the voltage boosting function.

Consuming the series or parallel blending of the SCs, DC sources, and power switches, SC-MLIs offer multilevel output voltage. SCs are charged to the appropriate voltage level by connecting them in parallel with the DC voltage source or directly through another charged SC. They then discharge their energy to the load in compliance with the voltage level criteria.

The productive and scalable integration of renewable energy sources is made achievable in large part by power electronics. A thorough review of power electronic technology supporting renewable energy systems is given by Tang et al. [1], who place particular emphasis on the continued development of high-efficiency converters and inverters. Similar to this, Zhu et al. [2] describe cutting-edge storage technologies for grid applications and emphasize the significance of energy storage systems in reducing the unpredictability of renewable sources. In their thorough evaluation of microcontroller-based battery management strategies in hybrid systems, Malele et al. [3] provide insights into new control mechanisms designed to ensure system efficiency and dependability. These are necessary for systems that integrate solar PV and storage to maintain energy balance. AI-based advanced control techniques have become increasingly popular. In order to improve performance in circumstances that are unpredictable, Younas et al. [4] suggest a deep learning-based stack LSTM controller for MPPT in a dual-stage PV system. Bana et al. [6] suggest an ANN-based current controller with superior MPPT for single-stage systems, while Kumar et al. [5] present a neural network with PNKLMF-based learning for grid-connected PV systems. In order to solve power prediction errors, Shafi et al. [7] use PV power forecasting for MPPT, while Hanzaei et al. [8] examine MPPT methods in light of environmental input variables. These contributions reveal how AI has begun to affect tracking algorithms. Eltamaly and Abdelaziz [9] provide an exhaustive textbook explanation of MPPT methods. For practical PV installations, Ali et al. [10] examine MPPT under both uniform and non-uniform irradiation. In their comprehensive studies of hybrid, optimum, and intelligent MPPT techniques, Bollipo et al. [11] and Yap et al. [12] support adaptive solutions. The practical application of newly developed MPPT algorithms is further explained by Kumar et al. [13], especially in situations with partial shading. While Joddumahanthi et al. [15] investigate the cuckoo search algorithm for MPPT, emphasizing its optimization potential, Hussain et al. [14] experimentally analyze ANN algorithms, reaffirming the viability of neural networks for dynamic irradiance situations. Multilevel inverters (MLIs) are essential for boosting PV systems' efficiency and power quality. In their study of switched-capacitor MLIs for PV applications, Gopal et al. [16] compare various topologies. In order to further enhance system compactness, Kumari et al. [18] further reduce the number of devices in five-level switched-capacitor MLIs. In order to reduce component complexity, Roy and Sadhu [19] recommended innovative step-up MLI designs that make use of switching capacitors. While Buduma et al. [17] concentrate on power coordination in DC microgrids leveraging hybrid energy storage, which is applicable to MLI-fed systems, Prabaharan and Palanisamy [21] combine boost converters with MLIs for more efficient operation. With a particular focus on power quality and reactive power management, Sun et al. [20] show how to employ ANNs to operate grid-tied residential PV systems. The study they conducted demonstrates how intelligent controllers promote the smooth integration of renewable energy sources into the overall framework.

The highlights of this research are given as follows:

- A five-level reduced switched-capacitor multilevel inverter (RSC-MLI) topology is proposed for solar PV integration, minimizing component count and enhancing voltage level generation.
- The system utilizes artificial neural network-based MPPT (ANN-MPPT) for accurate and dynamic maximum power tracking under varying solar irradiance conditions.
- ANN training is optimized using the Scaled Conjugate Gradient (SCG) algorithm, improving convergence speed and overall MPPT performance.
- The proposed inverter-MPPT combination achieves efficient power conversion with reduced total harmonic distortion (THD), improving output waveform quality.
- Comprehensive system modeling and performance evaluation are carried out using MATLAB/Simulink, validating the effectiveness of the design.

Designing and testing a five-level reduced switched-capacitor multilevel inverter (RSC-MLI) driven by a solar PV system is the primary purpose of this study. For effective power extraction, an ANN-based MPPT algorithm is utilized, and the SCG method is utilized for the best ANN weight tuning. The suggested solution seeks to improve MPPT performance, lower THD, and increase voltage boosting capacities. To ensure system efficiency and power quality, extensive simulations are carried out in MATLAB/Simulink.

The rest of the paper is organized as follows: Sect. 2 presents the PV system modeling circuit and equations; Sect. 3 presents the proposed five-level MLI configuration; results and discussion part was discussed in Sect. 4; and in Sect. 5, conclusions are presented followed by references.

2 System Modeling

(a) Grid-Connected PV System with ANN-MPPT

The grid-connected PV system was depicted in Fig. 1. It consists of various components like solar PV array, DC-DC boost converter, inverter, filter, and a grid. ANN-MPPT control was used to harvest maximum power from the solar PV array.

(b) PV System Modeling

Figure 1 shows the single diode equivalent model of solar PV system. It can be clearly observed that it consists of various components like current source, diode, and two series and shunt resistors. The open-circuit voltage can be taken as the output of the cell and I is the short-circuit current when both the load terminals are shorted.

The PV cell can be modeled using the below equations:

$$Photoncurrent, i_{ph} = i_d + i_{sh} + i \tag{1}$$

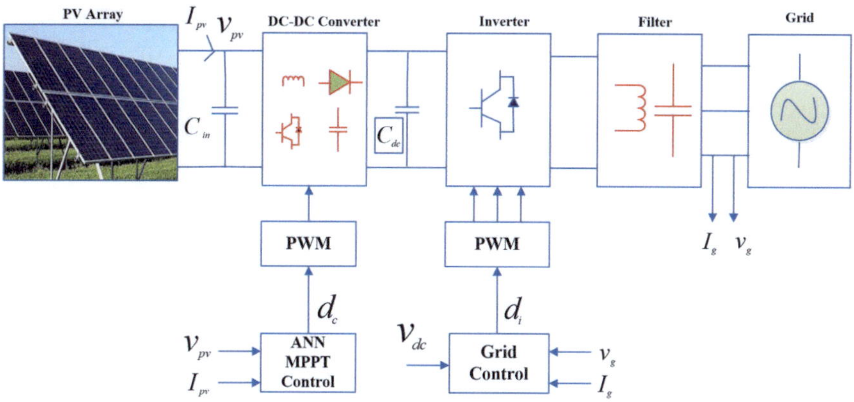

Fig. 1 Grid-connected PV system

$$Diodecurrent, i_d = i_0\left(e^{\left(\frac{v+i_{Rs}}{nV_t}\right)} - 1\right) \tag{2}$$

$$Terminalcurrent, i = i_{ph} - i_d - \left(\frac{v + i_{Rse}}{R_{sh}}\right) \tag{3}$$

$$Substitute equation\, 2\, in equation\, 3$$

$$i = i_{ph} - i_0\left(e^{\left(\frac{v+i_{Rs}}{nV_t}\right)} - 1\right) - \left(\frac{v + i_{Rse}}{R_{sh}}\right) \tag{4}$$

where i stands for the terminal current, i_0 for the reverse saturation current, i_{sh} for the current in the shunt resistance, and i_d for the diode current.

(c) DC-DC Boost Converter Modeling

Figure 3 shows the basic DC-DC boost converter and here switch plays a major role in the power transfer process. The modeling equations are described below:

$$\text{Output voltage}, V_0 = \frac{V_i}{1 - D} \tag{5}$$

$$Average\ value\ of\ inductor\ current,\ I_L = \frac{I_0}{1 - D} \tag{6}$$

$$Critical\ Inductance,\ L_C = \frac{D(1 - D)^2 * R}{2f} \tag{7}$$

$$Critical\ Capacitance,\ C_C = \frac{D}{2fR} \tag{8}$$

(d) **ANN Modeling**

Irradiance and temperature have been taken as the input datasets to train the ANN and the maximum voltage as the target one. The duty cycle which was generated using this ANN-MPPT control was then sent to a PID controller to generate required gate pulses to the switch present in the DC-DC boost converter (Fig. 4).

3 Proposed Topology

The proposed five-level MLI structure was depicted in Fig. 5. It merely consists of seven number of switches, a capacitor, and a DC voltage source. The switching scheme of the above configuration was shown in Table 2. The model output waveforms and the modes of operation of this MLI were shown in Fig. 2.

Mode 0: This mode refers to the zero output voltage. As stated in Table 1, in this mode, the switches S1, S2, S3, and S4 are ON while the rest of the switches will be in OFF position. The output voltage across the load will be a short-circuited voltage value and hence it is zero.

Mode 1: As stated in Table 1, in this mode, the switches S2, S3, S4, and S5 are ON while the rest of the switches will be in OFF position. The output voltage across the load will be equal to + 1 Vdc. The capacitor C charges to supply voltage.

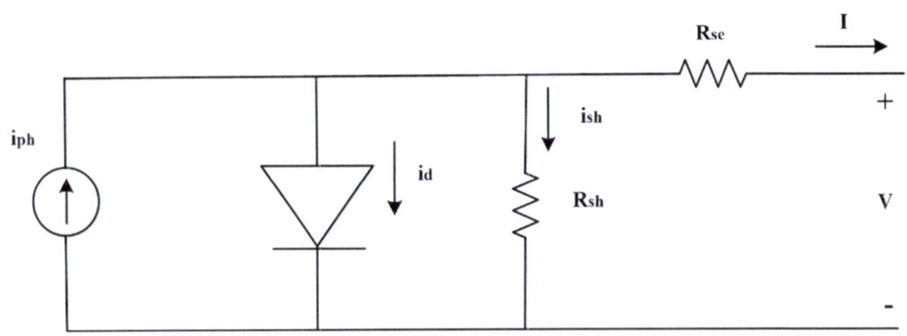

Fig. 2 Single diode equivalent model of solar PV system

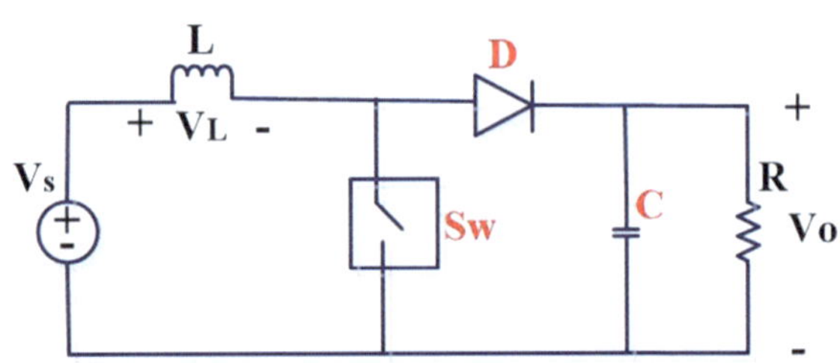

Fig. 3 DC-DC boost converter

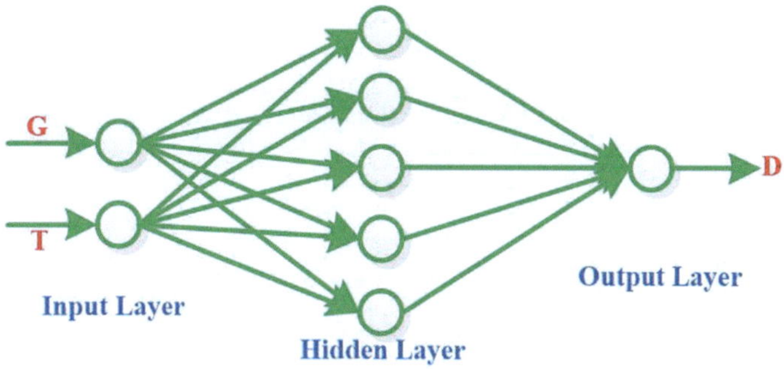

Fig. 4 ANN representation

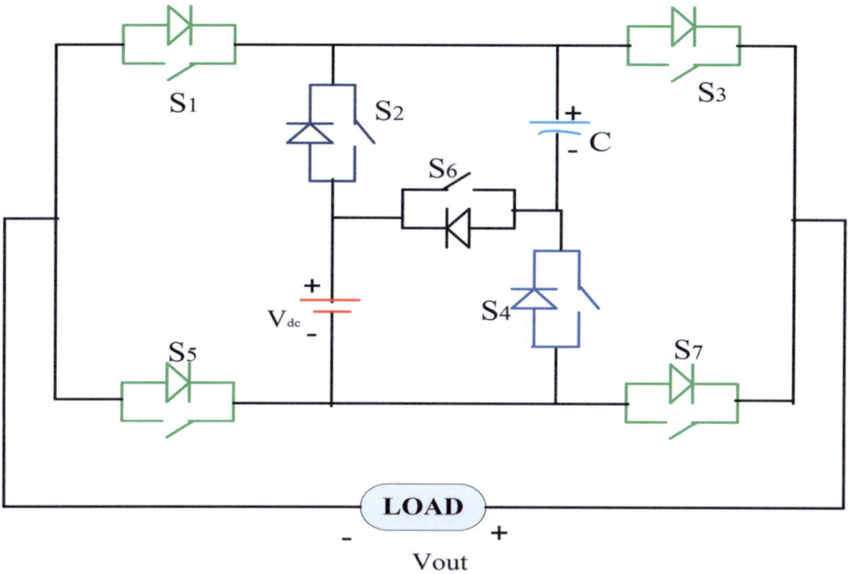

Fig. 5 Proposed MLI structure

Table 1 Switching scheme of proposed five-level MLI

Voltage levels	S1	S2	S3	S4	S5	S6	S7	C
+ 1Vdc	OFF	ON	ON	ON	ON	OFF	OFF	Charging
+ 2 Vdc	OFF	OFF	ON	OFF	ON	ON	OFF	Discharging
0	ON	ON	ON	ON	OFF	OFF	OFF	Charging
-1Vdc	ON	ON	OFF	ON	OFF	OFF	ON	-
-2Vdc	ON	OFF	OFF	OFF	OFF	ON	ON	Discharging

Table 2 Summary of author contributions

Ref. no.	Authors	Key focus area	Contribution
[1]	Tang et al. (2022)	Power electronics in renewable energy	Overview of enabling technologies for integration
[2]	Zhu et al. (2023)	Energy storage	Storage strategies for high VRE penetration
[3]	Malele et al. (2025)	Battery management	Microcontroller-driven hybrid energy system control
[4]	Younas et al. (2024)	Deep learning MPPT	Stack LSTM-based MPPT in dual-stage systems
[5]	Kumar et al. (2019)	Neural network MPPT	Hybrid HC and PNKLMF neural control
[6]	Bana et al. (2024)	ANN-based controller	Robust current control in grid-connected PV
[7]	Shafi et al. (2020)	Forecast-based MPPT	Predictive tracking using forecasted power
[8]	Hanzaei et al. (2020)	MPPT scheme review	Analysis based on irradiance and temperature inputs
[9]	Eltamaly& Abdelaziz (2020)	MPPT techniques book	Comprehensive MPPT algorithm coverage
[10]	Ali et al. (2020)	Irradiation conditions	MPPT under uniform and non-uniform shading
[11]	Bollipo et al. (2021)	MPPT taxonomy	Hybrid, intelligent, and classical methods
[12]	Yap et al. (2020)	AI-based MPPT	Neural network and fuzzy logic reviews
[13]	Kumar et al. (2023)	Shading effects	Emerging MPPT techniques under partial shading
[14]	Hussain et al. (2023)	ANN performance	ANN evaluation under dynamic irradiance
[15]	Joddumahanthi et al. (2025)	Cuckoo search MPPT	Nature-inspired optimization method
[16]	Gopal et al. (2025)	switched-capacitor MLI	PV-specific MLI topologies and assessments
[17]	Buduma et al. (2024)	Microgrid control	Hybrid storage coordination in DC microgrids
[18]	Kumari et al. (2023)	Reduced switch MLI	Five-level inverter with minimal switches
[19]	Roy & Sadhu (2021)	Step-up MLI	Switched-capacitor-based step-up inverter
[20]	Sun et al. (2017)	ANN for grid PV	Control and grid integration of residential PV
[21]	Prabaharan & Palanisamy (2016)	Boost + MLI	Integration with boost for efficient PV output

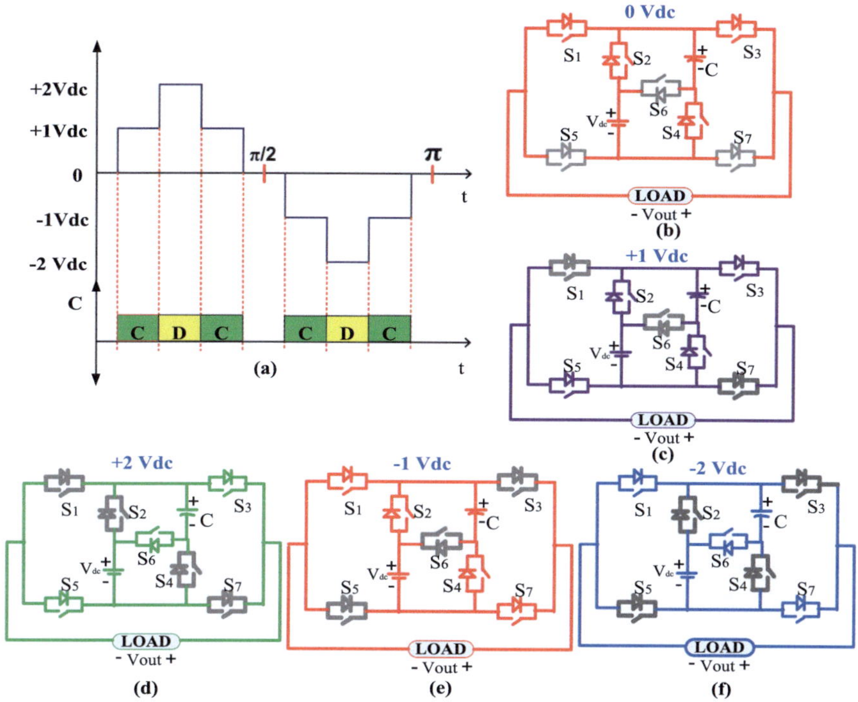

Fig. 6 **a** Five-level voltage waveform, schematic arrangement of **b** Mode 0, **c** Mode 1, **d** Mode 2, **e** Mode 3, and **f** Mode 4

Mode 2: As stated in Table 1, in this mode, the switches S3, S5, and S6 are ON while the rest of the switches will be in OFF position. The output voltage across the load will be equal to + 2 Vdc. The capacitor C will be in discharging mode.

Mode 3: As stated in Table 1, in this mode, the switches S1, S2, S4, and S7 are ON while the rest of the switches will be in OFF position. The output voltage across the load will be equal to -1 Vdc.

Mode 4: As stated in Table 1, in this mode, the switches S1, S6, and S7 are ON while the rest of the switches will be in OFF position. The output voltage across the load will be equal to -2 Vdc (Fig. 6).

4 Results and Discussions

The proposed five-Level SC-MLI has been simulated using MATLAB/Simulink software. The switching scheme was formulated in Table 1. The PV voltage and current were depicted in Fig. 7 and the PV output voltage obtained after implementing SCG-ANN-MPPT technique is approximately equal to 170 V as shown in Fig. 8.

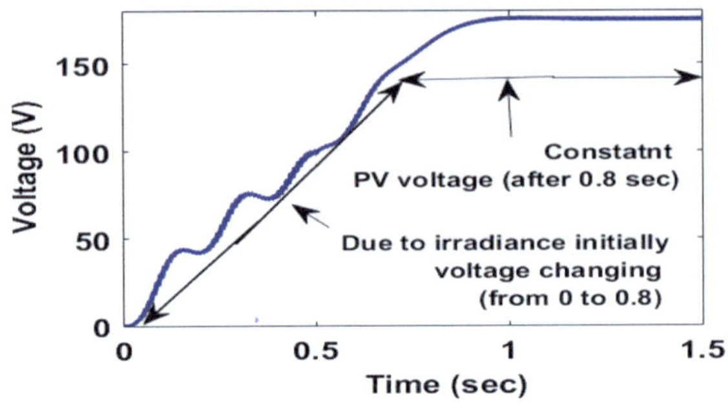

Fig. 7 PV output voltage

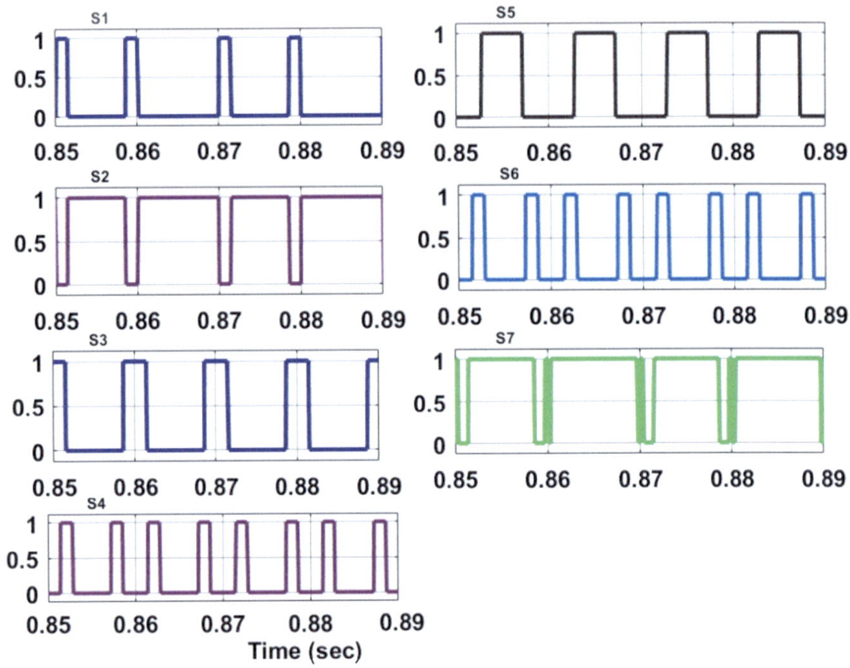

Fig. 8 Generation of pulses

Figure 7 illustrates the variation of photovoltaic (PV) voltage with time under dynamic change in the irradiance levels. It is observed that, initially, from 0 to 0.8 s, due to fluctuations in solar irradiance, the PV voltage increases nonlinearly. After 0.8 s, it is clearly shown that the PV voltage got stabilized and remains almost constant, indicating that the irradiance has reached a steady state. The switching pulses have been depicted in Fig. 8 and the duty cycle waveform that was generated after implementing ANN-MPPT is depicted in Fig. 9.

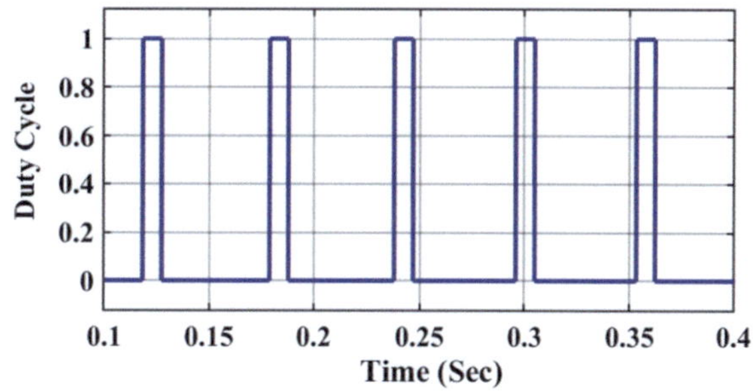

Fig. 9 Duty cycle waveform

As depicted in Fig. 10, the top graph represents a stepped output voltage waveform with a peak amplitude of $\pm$ 500 V which shows the characteristic behavior of a multilevel inverter. The bottom graph illustrates the corresponding load current waveform. It can be clearly observed that, initially, from 1.46 to 1.5 s, as the inverter is connected to a purely resistive (R) load, it results in a current waveform that is in phase with the voltage and exhibits a stepped profile. After 1.5 s, as the load changes to RL type, it results in making the current waveform to become smoother and phase-shifted relative to the voltage.

Figure 11 shows the output voltage waveform of the proposed multilevel inverter during both the transient and steady-state conditions. It is observed that the waveform exhibits transient oscillations initially from 0 to approximately 0.1 s. After this point, a neat stepped multilevel voltage waveform can be observed, which indicates that the inverter has reached steady-state operation.

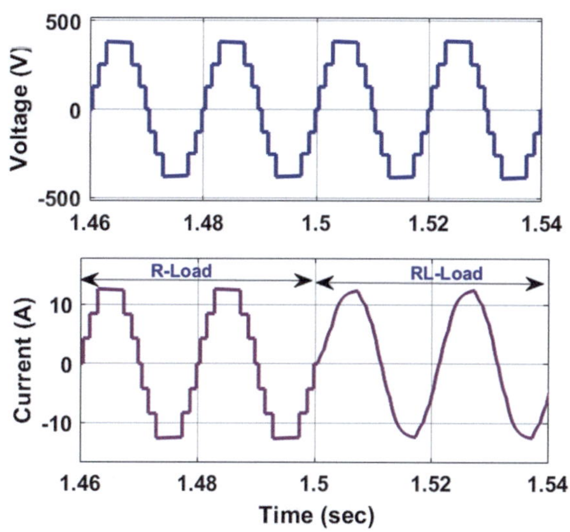

Fig. 10 Load voltage and current waveforms

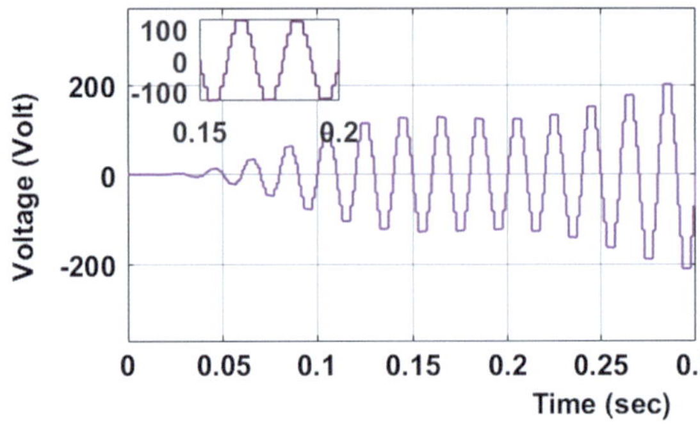

Fig. 11 Multilevel output voltage

5 Conclusions

In this research, an ANN-SCG MPPT controller has been accompanied with a five-level reduced switched-capacitor multilevel inverter (SC-MLI) and validated using MATLAB/Simulink. When temperature and irradiance shifts, the ANN trained using the scaled conjugate gradient algorithm quickly and effectively tracks the PV maximum power point. The recommended SC-MLI topology minimizes switching and conduction losses while producing a symmetric five-level staircase AC output with low THD value. Ultimately, the effective connection of a high-performance ANN-SCG MPPT with a small SC-MLI provides a scalable, efficient, and high-quality solution for solar PV system grid integration.

References

1. Z Tang Y Yang F Blaabjerg 2022 Power electronics: The enabling technology for renewable energy integration CSEE J Power Energy Syst 8 1 39 52 https://doi.org/10.17775/CSEEJPES.2021.02850
2. Yatindra Gopal, Kaibalya Prasad Panda, Pranda Prasanta Gupta, Dheeraj Kumar Dhaked, Yogendra Arya (2025) Development and performance assessment of switched-capacitor multilevel inverters for solar PV applications, Sustain Energy Technol Assess, 79, 104360, ISSN 2213–1388
3. RH Malele B Mendu BB Monchusi 2025 Microcontroller-driven battery management in hybrid energy systems: a systematic review of applications, control strategies, and emerging trends IEEE Access 13 29341 29360 https://doi.org/10.1109/ACCESS.2025.3539929
4. U Younas A AfsinKulaksiz Z Ali 2024 Deep learning stack LSTM based MPPT control of dual stage 100 kWp Grid-Tied solar PV system IEEE Access 12 77555 77574 https://doi.org/10.1109/ACCESS.2024.3407605
5. N Kumar B Singh BK Panigrahi 2019 PNKLMF-based neural network control and learning-based HC MPPT technique for multiobjective grid integrated solar PV based distributed generating system IEEE Trans Industr Inf 15 6 3732 3742 https://doi.org/10.1109/TII.2019.2901516
6. PR Bana S D'Arco M Amin 2024 ANN-based robust current controller for single-stage grid-connected PV with embedded improved MPPT scheme IEEE Access 12 100251 100262 https://doi.org/10.1109/ACCESS.2024.3429347

7. A Shafi H Sharadga S Hajimirza 2020 Design of optimal power point tracking controller using forecasted photovoltaic power and demand IEEE Trans Sustain Energy 11 3 1820 1828 https://doi.org/10.1109/TSTE.2019.2941862
8. SH Hanzaei SA Gorji M Ektesabi 2020 A scheme-based review of MPPT techniques with respect to input variables including solar irradiance and PV arrays' temperature IEEE Access 8 182229 182239 https://doi.org/10.1109/ACCESS.2020.3028580
9. AME Eltamaly AYE Abdelaziz 2020 Modern maximum power point tracking techniques for photovoltaic energy systems, Cham Springer Switzerland
10. A Ali 2020 Investigation of MPPT techniques under uniform and non-uniform solar irradiation condition–a retrospection IEEE Access 8 127368 127392 https://doi.org/10.1109/ACCESS.2020.3007710
11. RB Bollipo S Mikkili PK Bonthagorla 2021 Hybrid, optimal, intelligent and classical PV MPPT techniques: A review CSEE J Power Energy Syst 7 1 9 33 https://doi.org/10.17775/CSEEJPES.2019.02720
12. KY Yap CR Sarimuthu JM-Y Lim 2020 Artificial intelligence based MPPT techniques for solar power system: a review J Mod Power Syst Clean Energy 8 6 1043 1059 https://doi.org/10.35833/MPCE.2020.000159
13. M Kumar KP Panda JC Rosas-Caro A Valderrabano-Gonzalez G Panda 2023 Comprehensive review of conventional and emerging maximum power point tracking algorithms for uniformly and partially shaded solar photovoltaic systems IEEE Access 11 31778 31812 https://doi.org/10.1109/ACCESS.2023.3262502
14. MT Hussain A Sarwar M Tariq S Urooj A BaQais MA Hossain 2023 An evaluation of ANN algorithm performance for MPPT energy harvesting in solar PV systems Sustainability 15 11144 https://doi.org/10.3390/su151411144
15. Joddumahanthi V, Gopal Y, Pakalapati J, Kakinada VP, Potnuru V, Rajesh VDKSR (2025) Evaluation of cuckoo search MPPT algorithm for solar photovoltaic inverter applications. In: 2025 international conference on sustainable energy technologies and computational intelligence (SETCOM), Gandhinagar, India, pp 1–5, https://doi.org/10.1109/SETCOM64758.2025.10932350
16. Yatindra Gopal, Akanksha Kumari, Kaibalya Prasad Panda, Dheeraj Kumar Dhaked, Yogendra Arya (2025) Implementation and analysis of switched-capacitor multilevel inverters in solar photovoltaic system, Sustain Energy Technol Assess, 75, 104227, ISSN 2213−1388, https://doi.org/10.1016/j.seta.2025.104227
17. Buduma, Parusharamulu, Gopal, Yatindra, Kampara, Ravisankar, Kumar, Ranjeeth (2024) Power coordination and control of DC Microgrid with PV and hybrid energy storage system. Int J Emerg Electr Power Syst, https://doi.org/10.1515/ijeeps-2023-0328
18. Akanksha Kumari, Yatindra Gopal, Dheeraj Kumar Dhaked, Kaibalya Prasad Panda, Vijaya Kumar YN (2023) A single source five-level switched-capacitor based multilevel inverter with reduced device count, e-Prime—Advances in Electrical Engineering, Electronics and Energy, 5, 100235, ISSN 2772–6711
19. Yatindra Gopal, Yarrem Narasimhulu Vijaya Kumar, Akanksha Kumari, Om Prakash, Subrata Chowdhury, Abdullah A Almehizia (2023) Reduced device count for self balancing switched-capacitor multilevel inverter integration with renewable energy source" Sustainability 15(10) 8000. https://doi.org/10.3390/su15108000
20. Yatindra Gopal, Kaibalya Prasad Panda, Dinesh Birla (2022) Genetic algorithm and anti-predatory swarm optimisation-based solutions for selective harmonic elimination in multilevel inverters. IJPE, 15(3–4). https://doi.org/10.1504/IJPELEC.2022.122408
21. N Prabaharan K Palanisamy 2016 Analysis and integration of multilevel inverter configuration with boost converters in a photovoltaic system Energy Convers Manag 128 327 342

High Voltage Gain Multilevel Inverter Topology with Reduced Device Count and Low Voltage Stress

Prayag Chandra Barik, Narayan Nayak, Sidharth Samantara, Ashwini Kumar Sahu, Kasinath Jena, Dhananjay Kumar, Vishal Rathore, and Ramjee Prasad Gupta

Abstract Self-balanced switched-capacitor (SC) multilevel inverters (MLIs) have gained significant attention for their capability to enhance power quality in renewable energy and high-frequency electrical systems. This article presents a novel 13-level SCMLI topology (SCMLIT) that utilizes a reduced number of components (RDC) and operates with a single DC source. The proposed design achieves both voltage boosting and self-balancing of SCs through strategically managed charging and discharging sequences. The performance of the 13-level SCMLIT is validated through a comparative analysis against recent MLI topologies, considering parameters such as component count, total standing voltage, voltage gain, and cost efficiency. Furthermore, experimental results confirm the proposed SCMLIT's reliable operation under both linear and nonlinear load conditions, as well as during transient states.

Keywords Cost function · Multilevel inverter · Reduced device count · Switched capacitor · Voltage boost

P. C. Barik · N. Nayak · S. Samantara · A. K. Sahu · K. Jena
Department of Electrical and Electronic Engineering, ARKA Jain University, Jamshedpur, India

D. Kumar (✉)
Department of Electrical Engineering, Government Engineering College Siwan, Bihar, India
e-mail: dhananjaymanit23@gmail.com

V. Rathore
Department of Electrical Engineering, Maulana Azad National Institute of Technology, Bhopal, Madhya Pradesh, India

R. P. Gupta
Department of Electrical Engineering, MIT Muzaffarpur, Bihar, India

Y. Gopal et al. (eds.), *Multilevel Inverter Technology for Renewable Energy Systems*, Green Energy and Technology, https://doi.org/10.1007/978-3-032-03589-9_5

1 Introduction

In recent years, extensive research has been conducted on MLIs for a wide variety of applications requiring low, medium, and high levels of power [1]. The merits of MLIs, such as improved waveforms, reduced switching stresses, low distortion, and high modularity, have led to their greater popularity in industrial uses, electric vehicles, high-frequency alternating current distribution, and renewable energy generation. Although they both have drawbacks when it comes to voltage balancing of capacitors and complicated clamping circuits [2, 3]. The simple and adaptable design of the H-Bridge (HB) has made it well known in the academic community [4]. Diode-clamped MLI and flying capacitor MLI are two examples of traditional single DC source MLI architectures. The significant number of power semiconductor switches (*PSS*) is the primary drawback of these designs. This issue is most noticeable at very high output voltages.

They contribute to a rise in switching loss, size, and expense. All of these traditional MLIs have the same shortcoming: they can't provide a voltage-boosting ability. Numerous research efforts have been made to design small, inexpensive MLIs using switched capacitors (*SCs*) to address the aforementioned issues.

The modular arrangements shown in [5–7] allow for the development of larger amounts of output voltage. In order to provide an alternating current (AC) output at the load, these designs necessitate an H-bridge that may be reversed polarity. The outcome is an increase in total standing voltage due to the H-bridge switches having to withstand the maximum voltage stress. Another common form of topology is made by sandwiching SCMLI circuit among two half-bridges; this design is utilized to overcome the drawbacks of the aforementioned topologies. Step-up MLI architectures have been suggested by researchers [8], which allow for many capacitors to be charged and discharged at the same time. Nevertheless, as the output voltage increases, the total system voltage (TSV) escalates considerably, and the design requires a substantial quantity of bidirectional switches to optimize the circuit's charging capabilities. To get maximum levels from a pair of DC sources, the authors of [8, 9] described modules that made use of low semiconductors with inherently negative levels. Moreover, two 13-level SC topologies shown in [10–12] only need a DC source, although the voltage stress on some switches is still quite high.

In addition, authors in [13] describe a 13-level SC structure that yields the least gain. However, another topology attempts to boost high voltage (HV) but fails to maintain low switching stress [14]. The preferred topologies in [15–17] maintain all the benefits of the SC MLI and are able to synthesize 13 levels with voltage boosting under RL-type loads with RDC. Since the traditional HB is not present, less voltage stress is guaranteed in these circuits. The aforementioned literature survey provides possibilities for expanding a new topology that exhibits the following desirable properties:

1. Ability to boost the input voltage.
2. There is no need for manual regulation of capacitor voltages.
3. It requires less switching components.

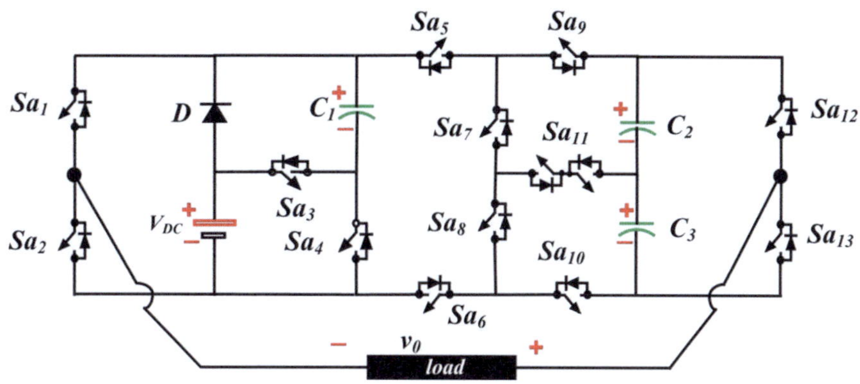

Fig. 1 Proposed 13-level SCMLIT

4. Switches are subjected to low voltage stress.
5. The inverter is versatile since it can accommodate a wide variety of power factors.
6. The topology has a very good response to dynamic changes.

2 Proposed 13-Level SCMLIT

2.1 Circuit Description of 13-Level SCMLIT

A diagrammatic representation of the suggested13-level SCMLIT is illustrated in Fig. 1.The design consists of 13 switches (Sa_1, Sa_2..Sa_{13}), 1 diode (D), and 3 capacitors (C_1, C_2, C_3). Among the 13 power switches, only S_{11} is a bidirectional switch. The topology is capable of producing 13 diagrammatic representation of the suggested 13-level SCMLIT which is illustrated in Fig. 1. The design consists of 13 switches (Sa_1, Sa_2..Sa_{13}), 1 diode (D), and 3 capacitors (C_1, C_2, C_3). Among the 13 power switches, only S_{11} is a bidirectional switch. The topology is capable of producing 13 different voltage levels, including $\pm 6V_{DC}$, $\pm 5V_{DC}$, $\pm 4V_{DC}$, $\pm 3V_{DC}$, $\pm 2V_{DC}$, $\pm 1V_{DC}$, and 0.

2.2 Principle of Operation

Table 1 and Fig. 2 show that the proposed topology's functioning may be characterized using 13 separate states, each defined by a different combination of switch positions, leading to 13 different voltage levels. The possible switching configurations for the suggested SCMLIT are shown in Table 1, where "0" indicates that the switches are OFF and "1" indicates that they are ON. The symbols "Δ," "∇," and "-" are used to denote the charging, discharging, and idle states of the capacitors, in that order.

State 1. $v_0 = 0$

In this operating scenario, $v_0 = 0$ can be achieved under this operating condition by simultaneously activating the switches $S_2, S_4, S_6, S_{10}, S_{13}$ or alternatively $S_1, S_4, S_5, S_9, S_{12}$. In this scenario, capacitor C_1 is charged to a voltage level of V_{DC}. The equivalent circuit corresponding to this operational state is depicted in Fig. 2 (a–b).

State 2. $v_0 = +1V_{DC}$

In this operating scenario, $v_0 = +1V_{DC}$ is achieved by simultaneously turning on switches $S_2, S_4, S_5, S_9, S_{12}$. During this mode, capacitor C_1 remains charged to a voltage of $ V_{DC}. The equivalent circuit representation corresponding to this state is shown in Fig. 2c.

State 3. $v_0 = +2V_{DC}$

By turning on switches $S_2, S_3, S_5, S_6, S_8, S_9, S_{11}, S_{12}$ all at once, the desired output voltage $v_0 = +2V_{DC}$ can be achieved under these operating conditions. Capacitor $2V_{DC}$ is charged during this mode using C_2. The equivalent circuit representation corresponding to this state is shown in Fig. 2d.

Table 1 Valid topology switching combinations

State	S_1	S_2	S_3	S_4	S_5	S_6	S_7	S_8	S_9	S_{10}	S_{11}	S_{12}	S_{13}	C_1	C_2	C_3	v_0
1	0	1	0	1	0	1	0	0	0	1	0	0	1	Δ	-	-	0
	1	0	0	1	1	0	0	0	1	0	0	1	0	Δ	-	-	0
2	0	1	0	1	1	0	0	0	1	0	0	1	0	Δ	-	-	$+1V_{DC}$
3	0	1	1	0	1	1	0	1	1	0	1	1	0	∇	Δ	-	$+2V_{DC}$
4	0	1	0	1	1	0	1	0	0	0	1	1	0	Δ	∇		$+3V_{DC}$
5	0	1	1	0	1	1	1	0	0	1	1	1	0	∇	∇	Δ	$+4V_{DC}$
6	0	1	0	1	1	0	1	1	0	1	0	1	0	Δ	∇	∇	$+5V_{DC}$
7	0	1	1	0	1	0	1	1	0	1	0	1	0	∇	∇	∇	$+6V_{DC}$
8	1	0	1	0	1	1	1	0	0	1	1	0	1	Δ	-	-	$-1V_{DC}$
9	1	0	0	1	0	1	0	1	0	0	1	0	1	Δ	∇	Δ	$-2V_{DC}$
10	1	0	1	0	1	1	0	1	1	0	1	0	1	Δ	∇	Δ	$-3V_{DC}$
11	1	0	1	0	1	1	0	0	1	0	1	0	1	Δ	∇	∇	$-4V_{DC}$
12	1	0	0	1	0	1	1	1	1	0	0	0	1	Δ	∇	∇	$-5V_{DC}$
13	1	0	1	0	0	1	1	1	1	0	0	0	1	∇	∇	∇	$-6V_{DC}$

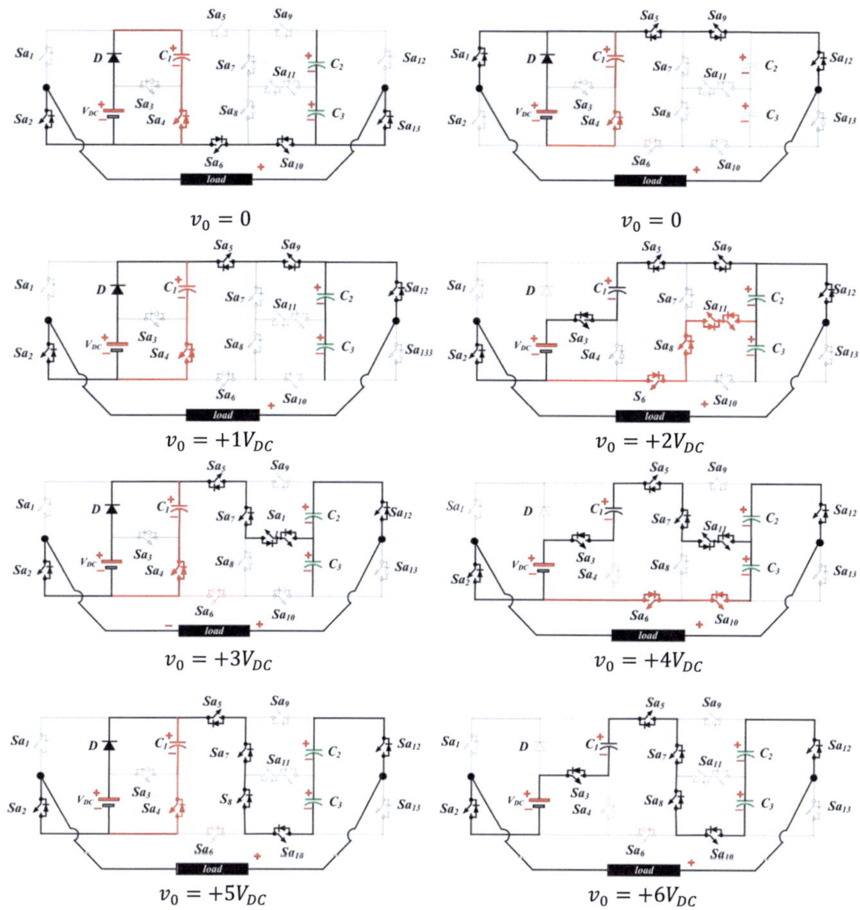

Fig. 2 Positive half-cycle operating state

State 4. $v_0 = +3V_{DC}$

In this operational condition, $V_0 = +3V_{DC}$ can be achieved by simultaneously switching S_2, S_4, S_5, S_7, S_{11},S_{12}. The load received the energy that was stored in capacitor C_2, while capacitor C_1 was charged to $1V_{DC}$. The model depicted in Fig. 2e is comparable to the one used at this level.

State 5. $v_0 = +4V_{DC}$

In this operational condition, $V_0 = +4V_{DC}$ can be reached by turning on switches S_2, S_3, S_5, S_6, S_7,$S_{10}S_{11}$, S_{12}. During this state, capacitor C_3 is charged to $2V_{DC}$, while capacitors C_1 and C_2 deliver their stored energy to the output. The equivalent circuit representation corresponding to this state is shown in Fig. 2f.

State 6. $v_0 = +5V_{DC}$

In this operational condition, $V_0 = +5V_{DC}$ can be attained by activating switches $S_2, S_4, S_5, S_7, S_8, S_{10}, S_{12}$. In this configuration, capacitor C_1 is charged to V_{DC}, while capacitors C_2 and C_3 release their stored energy to the load. The equivalent circuit representation corresponding to this state is shown in Fig. 2g.

State 7. $v_0 = +6V_{DC}$

In this operational condition, the output voltage $V_0 = +6V_{DC}$ can be reached by turning on switches $S_2, S_3, S_5, S_7, S_8, S_{10}, S_{12}$.In this mode, capacitors C_1, C_2, and C_3 discharge their stored energy into the load. The equivalent circuit representation corresponding to this state is shown in Fig. 2h.

Negative half-cycle voltage levels can be summarized using a similar approach, following the same principles and switching logic as applied to the positive voltage levels.

2.3 Voltage-Balancing Mechanism of Capacitors

The process for balancing the voltage on capacitors is covered in this section. The capacitors C_1, C_2, and C_3, integral components of the proposed SCMLIT, are linked in parallel with the input DC source for charging. During the generation of voltage levels 0, $\pm 1V_{DC}$,$\pm 3V_{DC}$,$\pm 5V_{DC}$, capacitor C_1 is charged to V_{DC}. Capacitors C_2 and C_3 are charged to $2V_{DC}$ during the generation of$\pm 2V_{DC}$,$\pm 4V_{DC}$ as detailed in Fig. 2c and d, respectively. At these stages, C_2 and C_3 are charged by connecting the DC source and C_1 in series. Capacitor C_1 discharges into the load during voltage levels$\pm 2V_{DC}$,$\pm 4V_{DC}$,$\pm 6V_{DC}$. During the creation of$\pm 3V_{DC}$,$\pm 4V_{DC}$,$\pm 5V_{DC}$,$\pm 6V_{DC}$, C_2, and C_3 also release the energy they have stored to the load. The operation of all capacitors is summarized in Table 1 and illustrated in Fig. 2. The series–parallel charging and discharging scheme inherently maintains voltage self-balancing across the capacitors, resulting in minimal voltage ripple [5, 6, 8].

2.4 Selection of Capacitor Size

SCMLIs rely heavily on capacitor capacitance because inverter performance is greatly impacted by voltage ripple across the capacitor. Less voltage ripple improves efficiency while more ripple increases losses. The discharge characteristics of the capacitor are also affected by three primary factors: the peak load current ($I_{max.}$), the maximum discharge time (t_m–t_n), and the power factor angle (∅) [5].

Therefore, the amount of charge discharged from the capacitor (ΔQc) can be expressed using Eq. (1).

$$\Delta Q_{c,i} = \int_{t_m}^{t_n} I_{max.} \sin(2\pi f t - \emptyset) dt \tag{1}$$

The longest discharging periods for capacitors C_1, C_2, and C_3 are, respectively, (t_1, t_{10}), (t_2, t_9), and (t_1, t_{10}).

For the suggested 13-level topology, $t_1 t_{2,\ldots\ldots} t_{10}$ are computed using Fig. 3 as follows [8]:

$$\begin{aligned} &t_1 = \frac{\sin^{-1}(1/6)}{2\pi f};\ t_2 = \frac{\sin^{-1}(2/6)}{2\pi f};\ t_3 = \frac{\sin^{-1}(3/6)}{2\pi f};\ t_4 = \frac{\sin^{-1}(4/6)}{2\pi f};\ t_5 = \frac{\sin^{-1}(5/6)}{2\pi f} \\ &t_6 = \frac{\pi - \sin^{-1}(5/6)}{2\pi f};\ t_7 = \frac{\pi - \sin^{-1}(4/6)}{2\pi f};\ t_8 = \frac{\pi - \sin^{-1}(3/6)}{2\pi f};\ t_9 = \frac{\pi - \sin^{-1}(2/6)}{2\pi f}; \\ &t_{10} = \frac{\pi - \sin^{-1}(1/6)}{2\pi f} \end{aligned} \tag{2}$$

Assuming p represents the maximum allowable voltage ripple (expressed as a percentage), the optimal capacitance value can then be determined using Eq. (3).

$$C_{i,opt} \geq \frac{\Delta Q_{C,i}}{P \times V_{DC}} \tag{3}$$

Under resistive load conditions (R_l), the optimum capacitance value can be derived from Eq. (2) and expressed as Eq. (4).

$$C_{i,opt} \geq \frac{8}{2\pi \times f \times R_{l\times} P} \tag{4}$$

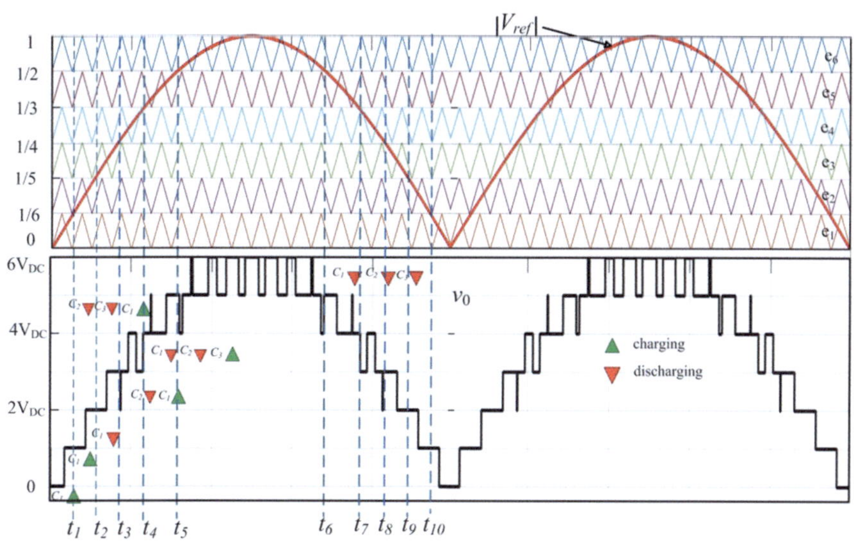

Fig. 3 Discharging instant of capacitance

For resistive-inductive load conditions, the optimal capacitance value can be determined by solving Eq. (3) and is expressed as follows:

$$C_{i,opt} \geq \frac{I_{max}}{2\pi \times f \times V_{DC} \times P}[\cos(0.8481 - \emptyset) - \sin\emptyset] \quad (5)$$

3 Results and Discussion

(A) Simulation Results

The suggested 13-level SCMLIT was modeled using the MATLAB/Simulink tools with the settings given in Table 2. A theoretical analysis was conducted on the SCMLIT dynamic performance. The waveform of the output is indicative of a resistive-inductive load configuration, as illustrated in Fig. 4. The supply voltage was found to be six times lower than the peak load voltage. Figure 4b shows that v_0 are unaffected by the abrupt change in the load. The voltages across capacitors C_1 and $C_2 = C_3$ did not deviate much from the 40 V to 80 V range, albeit there were small variations. Taken together, these results show that the suggested topology reliably operates under dynamic operating situations and successfully maintains the voltages of the switched capacitors independent of power factor and modulation index (M).

A laboratory prototype was constructed to assess the feasibility of the proposed SCMLIT design and carry out the required experiments. A dSPACE DS1104 real-time controller was employed to generate the gate signals.

The experimental setup of proposed 13-level SCMLIT is detailed in Fig. 5. And the experimental setting parameters are listed in Table 2. The experimental results for both the steady and transient states are displayed in Fig. 6. The steady-state analysis of the suggested 13-level SCMLIT is shown in Fig. 6a. In this case, an RL load is attached to the module. In this case, the output current takes the form of a sine wave, while the voltage level stays the same throughout the process. At a steady state, the

Table 2 Simulation and experimentation parameters for 13-level SCMLIT

Parameter	Specification
Supply voltage V_{DC}	10 V,20 V
Supply frequency f	50Hz
Carrier frequency	2kHz
Capacitor(C_1,$C_2 = C_3$)	2200 μF, 4700 μF
Load	30 + j25
Index of modulation (M)	0.95, 0.75, 0.5
IGBTs	FGW30XS65
Diode	TYN612 12A 600 V
Driver	TLP250

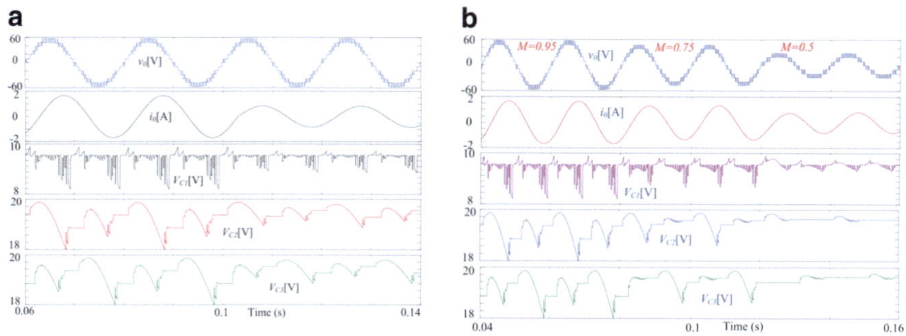

Fig. 4 Simulation outcomes: **a** Load change, **b** Modulation index change

voltage on the SCs remains relatively constant at approximately 10 V (for C_1) and 20 V (for C_2 andC_3), respectively.

The prototype is able to function normally despite a step change in load applied as shown in Fig. 6b. It has been revealed that the suggested 13-level SCMLIT generates 13 level with a boosting of 6X. In addition, the capacitors' voltages are automatically balanced when they are operated at their nominal voltage. This indicates that the V_{C1} is around 10 V, and V_{C2} and V_{C3} are approximately twice the supply voltage with very few voltage ripples.

The prototype can function with different values of M (0.95, 0.75, and 0.5), as shown in Fig. 6c. Modifications in M from 0.95 to 0.75 and from 0.75 to 0.5 decrease the output levels from 13 to 9 and from 9 to 7, respectively. The input voltage is set to double from 10 to 20 V in Fig. 6d. Accordingly, the inverter responds quickly

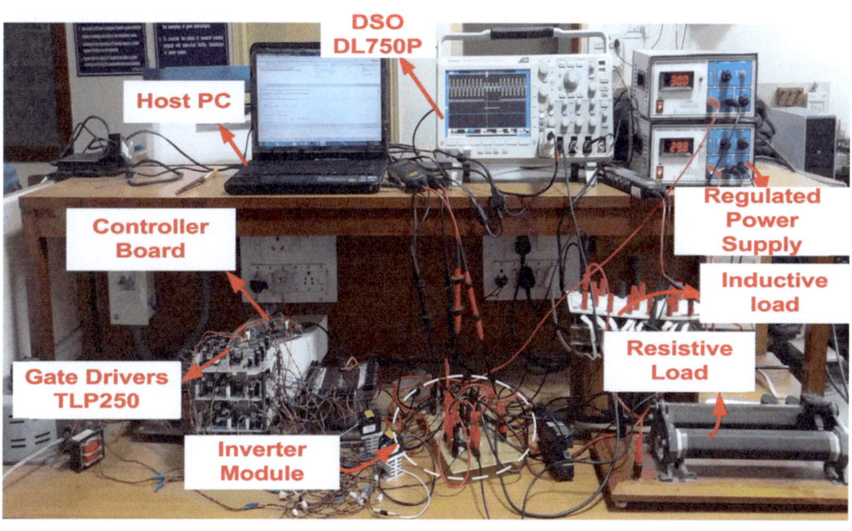

Fig. 5 Experimental setup of proposed 13-level SCMLIT [18]

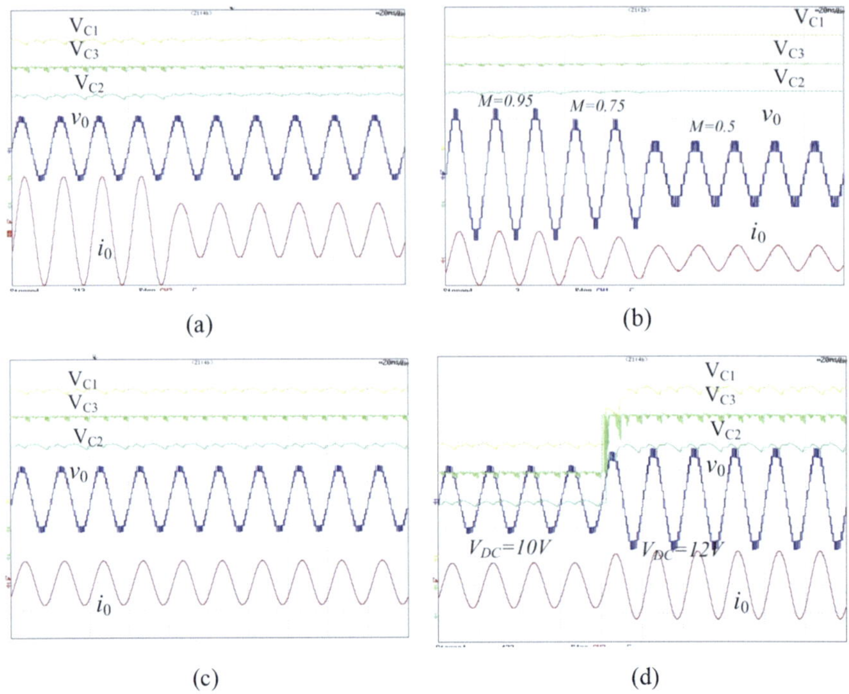

Fig. 6 Experimental outcomes **a** constant load, **b** varying load, **c** varying MI, **d** varying supply voltage

and generates13 level with peak values varying from 60 V to120V. Both the steady and transient responses of the suggested 13-level SCMLIT are excellent under the aforementioned conditions. Thus, theoretical analysis agrees with both the simulated and experimental results.

4 A Comparison Study of the SCMLIT

To evaluate the advantages and disadvantages of the proposed SCMLIT, it was compared with various contemporary topologies featuring a single-input DC source, focusing especially on capacitors (N_c), diode (N_d), drivers (N_{dri}), gain (g), MBV, switches (N_{sw}), and cost function (CF). Table 3 provides some quantitative comparisons. In the suggested 13-level SCMLIT, the PIV across any switch is less than $6V_{DC}$.The HB's switches are capable of withstanding the maximum load voltage in [5–7]. The definition of the cost function and per unit TSV of the inverter is as follows [11]:

Table 3 Comparative evaluation with the SCMLIs

Refs.	N_l	N_s	N_{sw}	N_c	N_d	N_{dri}	Gain (g)	$(TSV)_{pu}$	MBV	CF/g	
										$\alpha = 0.5$	$\alpha = 1.5$
[5]	13	1	19	5	–	19	6	6.5	6	0.59	0.67
[16]	13	1	14	4	–	14	1.5	12	2	1.948	2.564
[11]	13	1	13	3	2	13	6	6.5	6	0.480	0.522
[8]	13	1	29	5	5	29	6	5.66	1	0.944	0.980
[7]	13	1	10	5	10	10	6	9.83	6	0.51	0.577
[17]	13	1	12	3	4	12	6	6	3	0.435	0.51
[12]	13	1	10	4	4	10	6	6	6	0.397	0.512
[13]	13	1	14	3	1	13	3	6.33	2	0.876	1.038
[15]	13	1	14	3	1	14	6	5.5	3	0.445	0.615
[14]	13	1	13	3	2	13	6	5.5	3	0.432	0.503
[P]	13	1	14	3	1	13	6	5.8	4	0.434	0.508

$$CF = \frac{N_{sw} + N_d + N_c + N_{dri} + \alpha(TSV)_{pu}}{N_l} \times N_s \tag{11}$$

$$TSV_{pu} = \frac{\sum(TSV)}{V_m} \tag{13}$$

where α indicates the weight factor; the magnitude of this factor is related to the importance of switching components or TSV. Its values range from about 0.5 to 1.5. The suggested topology is less expensive to implement than the topologies proposed in [5] and [8], which rely on a larger number of semiconductor components. On the other side, the topologies [7] and [12] have RDC that are put under a greater degree of stress compared to the source voltage. Another important point to consider in the topologies described in [7, 8, 11, 12, 14, 17] requires a significant quantity of uncontrolled devices. In this configuration, the MBV is lower, but it CF/N_l in [8, 13–17]. As a result, the proposed 13-level SCMLIT offers significant benefits in terms of the overall cost.

5 Conclusion

The paper details a 13-level SCMLIT featuring RDC, sextuple boost, and self-voltage balancing. By contrasting the proposed 13-level SCMLIT with other current topologies, the benefits of RDC and lower costs are confirmed. The suggested 13-level SCMLIT was shown to be feasible by combining simulation and experimental data. The suggested 13-level SCMLIT is expected to be useful for HV systems that use low-DC input sources. For HV systems that use low-DC input sources, the proposed architecture is expected to be a viable option.

References

1. Gupta KK, Ranjan A, Bhatnagar P, Sahu LK, Jain S (2015) Multilevel inverter topologies with reduced device count: A review. IEEE Trans Power Electron 31(1):135–151
2. Kumari A, Gopal Y, Dhaked DK, Panda KP, Kumar YV (2023) A single source five-level switched-capacitor based multilevel inverter with reduced device count. E-Prime—Advances Electr Eng, Electron Energy 5:100235. ISSN 2772–6711
3. Meynard TA, Foch H (1992) Multilevel conversion: High voltage choppers and voltage-source inverters. In: Proc IEEE PESC, pp 397–403
4. Peng FZ, Lai JS, McKeever J, VanCoevering J (1995) A multilevel voltage-source inverter with separate DC sources for static var generation. In: Proc IEEE/IAS Annu Meeting, Orlando, FL, pp 2541–2548
5. Hinago Y, Koizumi H (2012) A switched-capacitor inverter using series/parallel conversion with inductive load. IEEE Trans Ind Electron 59(2):878–887
6. Babaei E, Gowgani SS (2014) Hybrid multilevel inverter using switched capacitor units. IEEE Trans Ind Electron 61(9):4614–4621
7. Gopal Y, Kumari A, Panda KP, Dhaked DK, Arya Y (2025) Implementation and analysis of switched-capacitor multilevel inverters in solar photovoltaic system. Sustain Energy Technol Assess 75:104227. ISSN 2213–1388, https://doi.org/10.1016/j.seta.2025.104227
8. Taghvaie A, Adabi J, Rezanejad M (2018) A self-balanced step-up multilevel inverter based on switched-capacitor structure. IEEE Trans Ind Electron 33(1):199–209
9. Roy T, Sadhu PK, Dasgupta A (2019) Cross-switched multilevel inverter using novel switched capacitor converters. IEEE Trans Industr Electron 66(11):8521–8532
10. Samadaei E, Kaviani M, Bertilsson K (2018) A 13-levels module (ktype) with two dc sources for multilevel inverters. IEEE Trans Ind Electron 66(7):5186–5196
11. Panda KP, Bana PR, Panda G A switched-capacitor self-balanced high-gain multilevel inverter employing a single DC source. IEEE Trans Circuits Syst II: Express Briefs (EarlyAccess). https://doi.org/10.1109/TCSII.2020.2975299
12. Ye Y, Chen S, Wang X, Cheng KWEE (2020) Self-balanced 13-level inverter based on switched-capacitor structure and hybrid pwmalgorithm. IEEE Trans Ind Electron
13. Bhatnagar P, Singh AK, Gupta KK, Siwakoti Y A switched-capacitors based 13-level inverter. IEEE Trans Power Electron. https://doi.org/10.1109/TPEL.2021.3098827
14. Sandeep N (2021) A 13-level switched-capacitor-based boosting inverter. IEEE Trans Cir Sys II: Exp Briefs 68(3):998–1002
15. Kim K-M, Han J-K, Moon G-W (2021) A high step-up switched-capacitor 13-level inverter with reduced number of switches. IEEE Trans Pow Elec 36(3):2505–2509
16. Zeng J, Lin W, Cen D, Liu J (2020) Novel K-type multilevel inverter with reduced components and self-balance. IEEE J Emg Sel T Pow Ele 8(4):4343–4354
17. Ye Y, Zhang G, Wang X, Yi Y, Cheng KWE (2022) Self-balanced switched-capacitor thirteen-level inverters with reduced capacitors count. IEEE Trans Ind Electron 69(1):1070–1076. https://doi.org/10.1109/tie.2021.3050378
18. Jena K, Dhal G, Gupta KK, Kumar D, Singh V, Sahu PK (2025) Switched capacitor-based 13-level multi-gain multilevel inverter with reduced switches and voltage stress. Electr Eng 107(4):4077–4090

Novel Switched Capacitor Self-Balanced Reduced Switch Multilevel Inverter for Photovoltaic System

Yatindra Gopal, J. Vijaychandra, Akanksha Kumari, Kaibalya Prasad Panda, and B. Parusharamulu

Abstract Reduced switch multilevel inverters (RS-MLI) are currently used in industry and for renewable energy (RE) resources. The most well-known renewable energy source for producing electricity is photovoltaic (PV) cell. To generate the sustainable energy from PV system maximum power point (MPP) technique is using in this paper based on Fuzzy Logic controller (FLC) technique in order to generate the maximum power. The FLC MPPT determines the reference PV voltage assurance to provide the best possible power transfer between the PV system and the load-connected RS-MLIs. This study implements a novel 7-level PV-based switched capacitor operation-based MLI (7L-PV-SCMLI) topology with the goal of minimizing the number of semiconductor devices and reduced total harmonic distortion (THD). To calculate the firing angles of 7L-PV-SCMLI, selected harmonic elimination-based anti-predatory particle swarm optimization (APSO) technique is used. MATLAB/Software version "9.10.0.1602886 (R2021a)" platform is used to validate the results performance.

Y. Gopal (✉) · J. Vijaychandra · B. Parusharamulu
Department of Electrical Engineering, National Institute of Technology, Jote, Arunachal Pradesh, India
e-mail: hellogopal4u@gmail.com

A. Kumari
Department of Electronics and Communication Engineering, Sreenivasa Institute of Technology and Management Studies, Chittoor, Andhra Pradesh, India

K. P. Panda
Department of Electrical Engineering, School of Technology, Pandit Deendayal Energy University (PDEU), Raisan, Gandhinagar, Gujarat, India

J. Vijaychandra
Department of Electrical and Electronics Engineering, Lendi Institute of Engineering and Technology, Viziangaram, India

B. Parusharamulu
Department of Electrical Engineering, Lendi Institute of Engineering and Technology, Viziangaram , India

Y. Gopal et al. (eds.), *Multilevel Inverter Technology for Renewable Energy Systems*, Green Energy and Technology, https://doi.org/10.1007/978-3-032-03589-9_6

Keywords 7L-PV-SCMLI topology · PV Renewable energy (RE) · Fuzzy Logic Controller (FLC) · Anti-predatory particle swarm optimization (APSO) · Total harmonic distortion (THD)

1 Introduction

The most widely used renewable energy source is PV due to its continuous energy supply, low maintenance costs, and cleanliness. PV panels' output voltage is greatly decreased in partial shadowing conditions, which lowers their efficiency. The PV array characteristic curve shows the ideal MPP value, which is chosen for the popular kind of maximum power point tracking technology (MPPT) in order to achieve constant irradiance [1–4]. Hot spots form in the PV strings as a result of bypass diodes, making the predicted MPPT technique unsuitable for determining the ideal maximum point. Many optimized techniques are described in the literature [5] to optimize the various values of maxima at the time of partial shading. The perturbation and observation (P&O) technique is typically utilized since it is easy to apply and effortless; nevertheless, under partial shade conduction, this procedure may fail or oscillate around the MPP. When certain aspects of the P&O technique are overcome, incremental conductance (IC) also shows up; however, the process requires more intricate calculations [6]. The rules-based FLC MPPT technique is used to the PV system in this paper. PV input voltage and current are used as inputs in the FLC-MMPT approach, and the resulting duty cycle is computed as the output. An MPPT controller is equipped with FLC MPPT to enhance the efficiency of energy exchange for DC–DC boost. Since most loads are AC, the DC–DC improved output voltage from the DC converter must be modified in the form of AC for commercial purposes. Various MLI topologies have been available and extensively studied for the conversion of DC to AC in recent research projects. Researchers have focused on MLIs in the context of solar PV system applications, wind power conversion, uninterruptible power supply, and AC motor drive systems [5].

For eliminates the electromagnetic interference (EMI), increased efficiency, and the ability to handle high voltage, are all considered from the perspective of reduced voltage stress on semiconductor switches [7–10]. Many different voltage steps are produced by the output voltage obtained from MLIs. The number of DC sources, DC-link capacitors, count switches, voltage steps, and output power quality are the primary determinants of MLI setup [11]. Neutral-clamped (NC) MLIs, which were first introduced in 1981, are the second type of MLI. The third type, flying-capacitor (FC)-based MLIs, was introduced in the early 1990s.

The cascaded H-bridge (CHB) MLI was first planned in 1975. CHB MLI topology is built on isolated, steady DC voltage sources that are connected to each H-Bridge cell at its input, making it appropriate for use with renewable energy sources. Asymmetrical configurations use various values for the number of DC sources, which might result in a more complex circuit, whereas symmetrical configurations assume an equivalent number of DC sources. When the number of H-Bridges in CHB MLI increases, so do the number of DC voltage sources; this is because the region has more DC sources and switches that are needed [12–15]. In this research, switching capacitor-based MLI techniques are used to sustain an inductive load with a high voltage gain, reduce the number of devices, and simplify the circuit.

Nowadays, determining the optimum feasible power superiority with improved efficiency by lowering THD and switching losses is the primary challenge facing MLIs. Low-frequency switching is one of the two fundamental types of management strategies, and it can gives obtainable concert than higher-frequency solution [13, 15, 16]. The predictable pulse width modulation (PWM), such sinusoidal PWM and vector modulation have high switching frequencies and speed up transient reactions [12].

Furthermore, DI carrier-based and carrier-less PWM approaches that effectively lower EMI and distortion are introduced [13]. Higher-order harmonics can be suppressed with low-pass filters, increasing MLI efficiency. Furthermore, the PWM approach cannot completely eliminate lower order harmonics like the third and fifth since they are leading in nature. Numerous optimization methods, including genetic algorithms (GA), harmony search strategies (HST), and partial swarm optimization (PSO), are employed in the literature to lower the THD. By solving transcendental nonlinear equations, the anti-predatory particle swarm optimization (APSO) technique is utilized in this paper to determine the ideal switching angle value in order to eliminate certain lower order harmonics.

2 Uniform and Non-uniform Irradiance Conduction of PV Characteristics

The production of PV sources is directly impacted by variations in temperature and radiation [17–19]. A distinct peak is produced by the power–voltage curve when the PV arrays are assembled under constant solar insolation. On the PV curve, however, a PV array with partial shading conduction produces a number of peaks, with one optimal point—known as the global maximum point, or GMP and another as the local maximum point, or LMP. Similar to reverse voltage biasing, the shaded cells receive reduced irradiance. Due to the steady flow of current in each series-structured module, the supplied current is comparable to that through the unshaded cell. The characteristic function of a photovoltaic panel in a standard series parallel configuration under uniform and partial shade conduction is depicted in Fig. 1.

A discrepancy in insulation between two modules may result in one of the modules in the string receiving less light as a result of shadowing conduction. As a result, the I–V curve displays two current waveforms. A single global peak (GP), shown by curve b, and several local peaks characterize the subsequent PV curve. Furthermore, the PV array only displays the single peak (curve 3) in Fig. 1. It can be challenging to choose the GMP, hence different MPPT techniques are utilized to increase the PV array output.

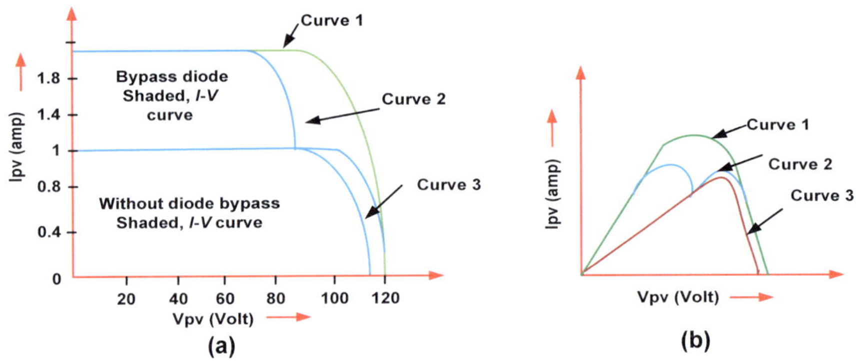

Fig. 1 I–V and PV curve **a** for condition of uniform insolation **b** for condition of partial shading

To get the best power out of a PV system, many researchers are working on MPPT approaches. The two main types of MPPT approaches that have been developed in recent decades are conventional and soft computing techniques. Hill climbing (HC), incremental conductance (IC), global MPPT, and fractional open and short circuit [20] are all used in conventional approaches. The optimal maximum power point (MPP) can be tracked using well-known techniques like P&O, HC, and IC in good conditions; however, these techniques break down in partial shading. Additionally, the steady-state oscillation of these methods is large, their tracking speed is slow, and their convergence is poor.

Consequently, methods based on soft computing have been established in the field of solar PV research to track MPP optimum data in partial shade conditions. For tracking the best MPPT value under partial shading conditions, several soft computing techniques are available, including genetic algorithms, artificial neural networks, cuckoo search, colony bee, fireflies, and random search techniques. The ideal data of MPPT is tracked in this paper using an FLC technique based on soft computing. There is no need for the FLC approach in the mathematical model [21].

Non-linear optimization issues can be solved with this robust rule-based method. Due to important benefits including dependable performance, a convenient user interface, ease of implementation, qualified validation, and duty cycle values that are ideal for DC–DC boost converters.

3 Boost Converter

To boost the PV voltage DC to DC converter is used between PV system and inverter. Because the generated power by PV system is not sufficient to fulfill the load demand. The Fuzzy MPPT technique and DC to DC boost converter is used in this paper. The inductor current does not drop to zero when the converter is operating in continuous conduction mode. The following equations mention the duty cycle, inductor (L), and capacitor (C):

$$\frac{V_{out}}{V_{in}} = \frac{1}{1-Q} \tag{1}$$

$$\text{Inductor (L)} = \frac{V_{in} \times Q \times T}{\Delta_{iL}} \tag{2}$$

$$\text{Capacitor (C)} = \frac{V_{out} \times Q}{R \times \Delta V_{out} \times f} \tag{3}$$

where V_0 is the boost converter's output voltage, V_{in} is its input voltage, and Q is its duty cycle. Additionally, capacitor voltage is represented by QV_0. When the Q value is between 0 and 1, the output exceeds the input voltage. Fuzzy MPPT and the PV–DC converter concept provide the best duty cycle values for the boost converter using PV. The PV system with DC to DC converter and FLC MPPT technology is depicted in Fig. 2.

It is observed that as the duty cycle increases, the PV system average current rises and its output voltage falls. As a result, increase the Q. The PV array V–I characteristic's left side is the operational point in that movement. Similarly, a decrease in Q results in a decrease in the average current and an increase in the average voltage of the PV array. The right PV array's operating point likewise shifted as a result of the shifting finding. Using the modified incremental conductance (IC) MPPT technique, this issue can be resolved by automatically adjusting the DC–DC converter Q value to produce the ideal DC voltage at the terminal. Uncertainties such as un-modeled physical quantities and irregular variations in operation point have contributed to the significant interest in IC based on soft computing technology (SCT) MPPT under uniform insolation. To determine the optimal MPP value and provide a suitable Q for the DC–DC converter, this article used the FLC MPPT technique.

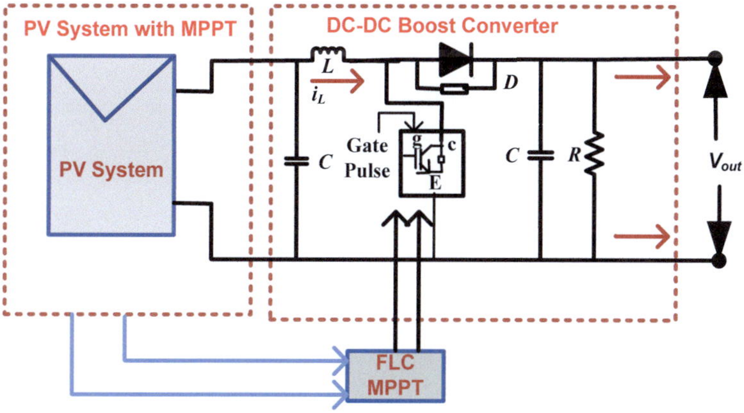

Fig. 2 Basic concepts of MPPT and DC–DC boost converter

The main benefit is that it solves nonlinear optimization problems with robustness and without the need for a system mathematical model.

4 MPPT-Based Fuzzy Logic Controller

The most recent literature review has taken into account a variety of MPPT techniques. These techniques are remarkable for their ability to enhance tracking speed under changing ambient conditions, reduce oscillation around the maximum power point, and respond quickly. It has been determined from the literature that several researchers have approached P&O, hill climbing (HC), IC, modified IC, and MPPT based on the soft computing technology (SCT) under uniform insolation. These SCT-based methods have been gaining a lot of attention because of uncertainties such as irregular variations in operating point and un-modeled physical quantities. The FLC MPPT technique is used in this work to determine the optimal MPP values and establish a suitable duty cycle for the DC–DC converter. This approach provides robustness to solve nonlinear optimization problems without the need for a mathematical model. The fundamental concept of an MPPT controller based on FLC is shown in Fig. 3. Figure 4 shows the distinct process construction of FLC.

The classification of FLC manly in four types such as *fuzzification set*, *knowledge base set*, *fuzzy interfaces engine*, and *defuzzification set*. Fuzzy implication and fuzzy rule conniving are used in these processes to determine the best system response.

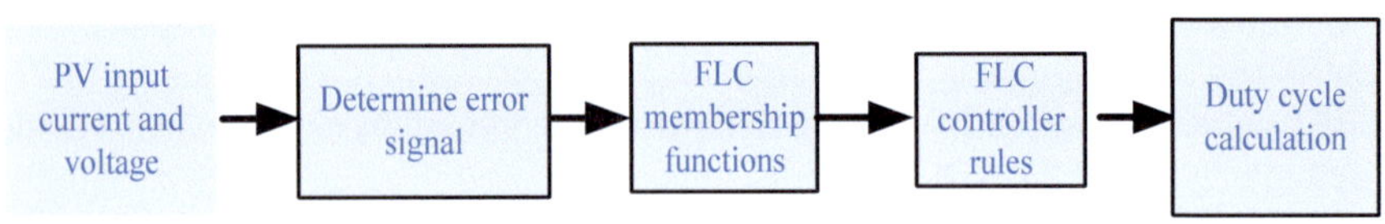

Fig. 3 Process of FLC

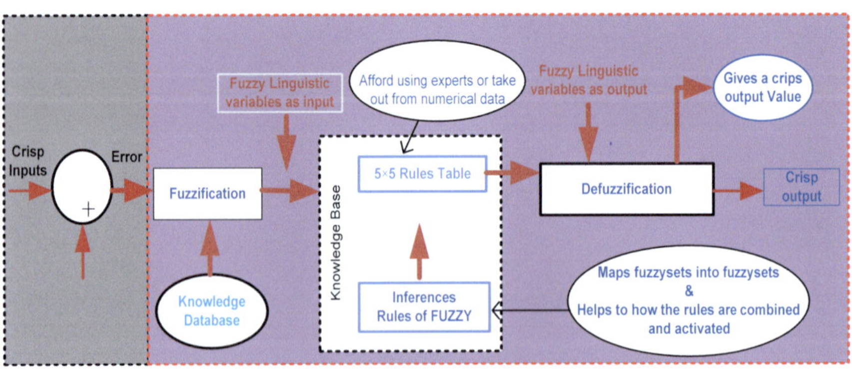

Fig. 4 Process structure of FLC

This section explains the error Z and fluctuating error ΔZ that are commonly found in FLC MPPT inputs. The roles of the four different FLC blocks are explained as follows.

Fuzzification is the process of shifting the range of values of input variables into the resulting communication universes by using a fuzzy set with linguistic labels to measure inputs value rules of control. Determine power continuously calculates the real voltage (V) and current (I) of the PV array. At sampling time (K), variables Z and CZ have the following expressions:

$$Z(K) = \frac{P(K) - P(K-1)}{I(k) - I(k-1)} \tag{4}$$

$$CZ(K) = Z(k) - Z(k-1) \tag{5}$$

Here, $P(K)$ and $I(K)$ of the PV array, respectively, define power and current. If the instant K input is $Z(K)$, the operational point is positioned on the left side of the MPP; if not, the $CZ(K)$ input error expresses the direct displacement of the operational point on the right side. The duty ratio of delta-D is altered using a DC–DC converter, similar to the output of the suggested controller. To determine the ideal point, slope $Z(K)$ is carried backwards to the point where slope is zero by varying the duty ratio value. This work considers input values between –1 and 1 in such a way that

Interface engine: This stipulates that output control to be accomplished through the composition process; for instance; the computer's MATLAB fuzzy toolbox establishes MAX–MIN and MAX-DOT. The MAX–MIN approach often gets used in this work. Every rule's membership is laid out using the MIN (minimum) and MAX (maximum) function outputs. The FLC rule table appears in Table 1. Five fuzzy subsets are used for allotting membership function values to linguistic variables: NH stands for consider (negative high), NS for consider (negative small), ZZ for consider (zero), PH for consider (positive high), and PS for consider (positive small). Figure 5 displays the membership function's outline, which can be adjusted to fit the best possible system, and the segmentation for fuzzy subsets. Table 1 represents the fuzzy control rules with associated memory.

The following mechanisms are typically employed in fuzzy control: Somme-Prod, Max-Prod, and Max–Min interface techniques. This is the Max–Min of fuzzy

Table 1 FLC control rules for associative memory

CE/E	NH	NS	ZZ	PS	PH
NB	ZZ	ZZ	NH	NB	NB
NS	ZZ	ZZ	NS	NS	NS
ZO	NS	ZZ	ZZ	ZZ	PS
PS	PS	PS	PS	ZZ	ZZ
PH	PH	PH	PH	ZZ	ZZ

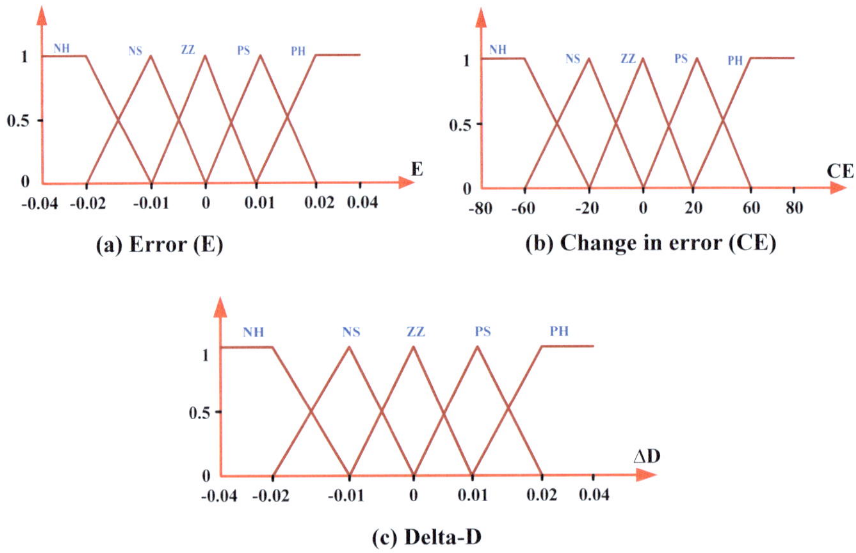

Fig. 5 Functions of membership for output and inputs

arrangement in the Mamdani inference approach that is explored in this paper. Figure 6 displays the 49 fuzzy control rules graph in three dimensions (3D).

Knowledge-base: Two components render the framework of knowledge base: fuzzy control (rule base) and fuzzy set (data base). The function of the fuzzy set is to adjust fuzzy data in an FLC and to reflect rules of fuzzy control. The logic underlying this process is based on experience, so it should be mentioned that the proper selection of a term set's membership functions is critical to an application's success.

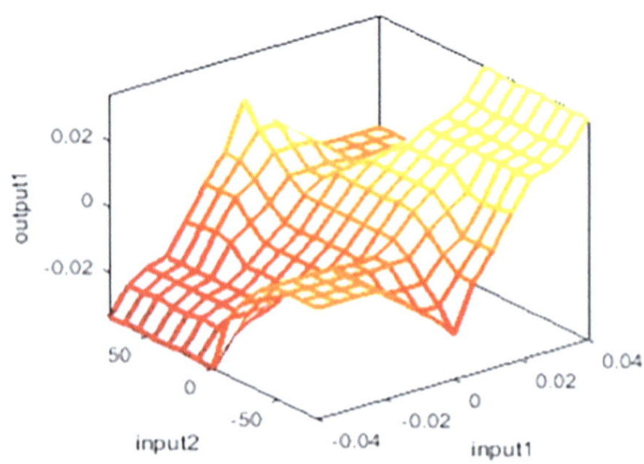

Fig. 6 FLC MPPT mesh surface

Defuzzification: Following rule forecasting, the FLC undertakes fuzzification to determine the fuzzy controller soft output. Employing a DC–DC converter as an accurate control signal, forecasting the defuzzification the transformation of fuzzy knowledge into reliable information is crucial to its access. The center of gravity technique is taken into consideration while performing defuzzification. The defuzzification method for the center of gravity in the system rules is defined by

$$\Delta Q = \frac{\sum_{c=1}^{n} w(\Delta Q_c).\Delta Q_c}{\sum_{c=1}^{n} w(\Delta Q_c)} \tag{6}$$

where w is the weighting factor and n is the maximum number of efficient rules. Equation (7) uses the defuzzification approach to obtain the fuzzy controller output, which is the change of duty ratio $\Delta Q(f)$, which is then scaled by the gain $SP_{\Delta D}$. Gain $SP_{\Delta Q}$ is used to determine the actual duty ratio, $Q(f)$.

$$Q(f) = Q(f-1) + SP_{\Delta Q} \times \Delta Q(f) \tag{7}$$

5 Proposed 7L PV-SCMLI

The 7L output voltage in this recommended design is generated by a single dc source. In addition, the 7L-PV-SCMLI design uses two capacitors (C1–C2), three diodes (D1–D3), and eight switches (S1–S8). Vo is the output voltage, while the PV voltage source serves as the MLIs input.

A. **Procedure of 7L SCMLI**

In order to generate a seven-level staircase output waveform, the components like Single PV dc source, two capacitors and eight switches have been taken. The steps output generation of 7L, six is bipolar levels and 0 level at the output voltage waveform. Figure 7 represent the PV based 7L-PV-SCMLI.

To generate the 7L of output voltage operation of different case for7L-PV-SCMLI as under:

Case 1 (V_{dc}): When switches S_1, S_2, S_3, S_7, and S_8 are off and switches S_4, S_5, and S_6 are in working mode, the voltage output level is $+V_{dc}$. Reverse **operation of diodes D_1, D_2, and D_3 as well as the conduction mode of capacitors C_1 and C_2.** **Case II** ($+2V_{dc}$): In this instance, switches S_1, S_4, S_7, and S_8 are not functioning, but switches S_2, S_3, S_5, and S_6 are in the ON stage. **Case III** ($+3V_{dc}$): When the switches S_2, S_3, S_4, S_7, S_8 is OFF and S_1, S_5, S_6 are ON stage, the $+\ 3V_{dc}$level of voltage can be obtained. **Case IV** ($-3V_{dc}$): When $-3V_{dc}$ switches are present, S_1, S_7, and S_8 are ON levels and S_2, S_3, S_4, S_5, and S_6 are OFF. Switches S_1, S_4, S_5, and S_6 are off in **Case V** ($-2V_{dc}$), but switches S_2, S_3, S_7, and S_8 are on. **Case VI** ($-V_{dc}$):In this scenario, switches S_1, S_2, S_3, S_5, and S_6 are off, while switches S_4, S_7, and S_8 are on.

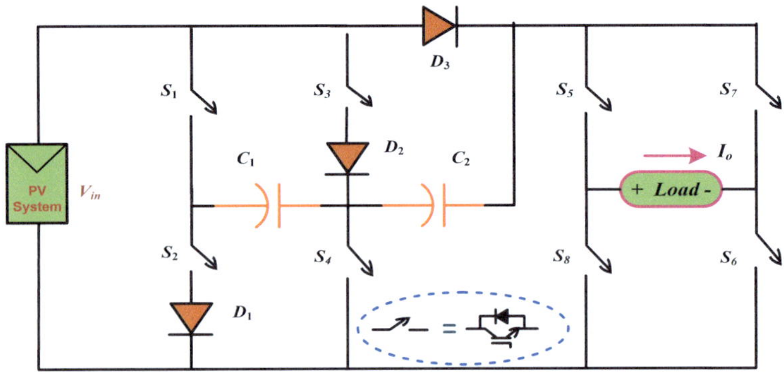

Fig. 7 PV-based 7L switched capacitor multilevel inverter (7L-PV-SCMLI)

Table 2 Switching states of the 7L-PV-SCMLI with capacitor charging (↑) and capacitor discharging (↓) period

Voltage stage	Working switches	Capacitor (C_1)↑	Capacitor (C_2)↓
0	S_5, S_7	–	–
1	S_4, S_5, S_6	-	↑
2	S_2, S_3, S_5, S_6	↑	↓
3	S_1, S_5, S_6	↓	↓
-3	S_1, S_7, S_8	↓	↓
-2	S_2, S_3, S_7, S_8	↑	↓
-1	S_4, S_7, S_8	–	↑

In **Case VII** ($0V_{dc}$), switches S_1, S_2, S_3, S_4, S_5, and S_7 are off; however, switches S_6 and S_8 are in conduction mode.

In the given Table 2, the switching scheme of the proposed topology and charging and discharging period of capacitors phenomenon has been provided.

Figure 8 illustrates the 7L-PV-SCMLI's produced voltage waveform. This technique is suitable for high-power converters because it employs several switches for every output voltage stage, raising the output voltage waveform's quality. A look-up table is used to organize the many sets of solutions for nonlinear transcendental equations.

6 Calculation of THD Using Optimization Algorithm

As discussed in Sect. 1, the output voltage of MLI is at present synthesized primarily through the selected harmonic elimination (SHE) technique. A quarter wave balanced output hits when the output wave for the negative half cycle is equal to the output wave for the positive half cycle. Figure 8 represent the non-equal dc source pattern

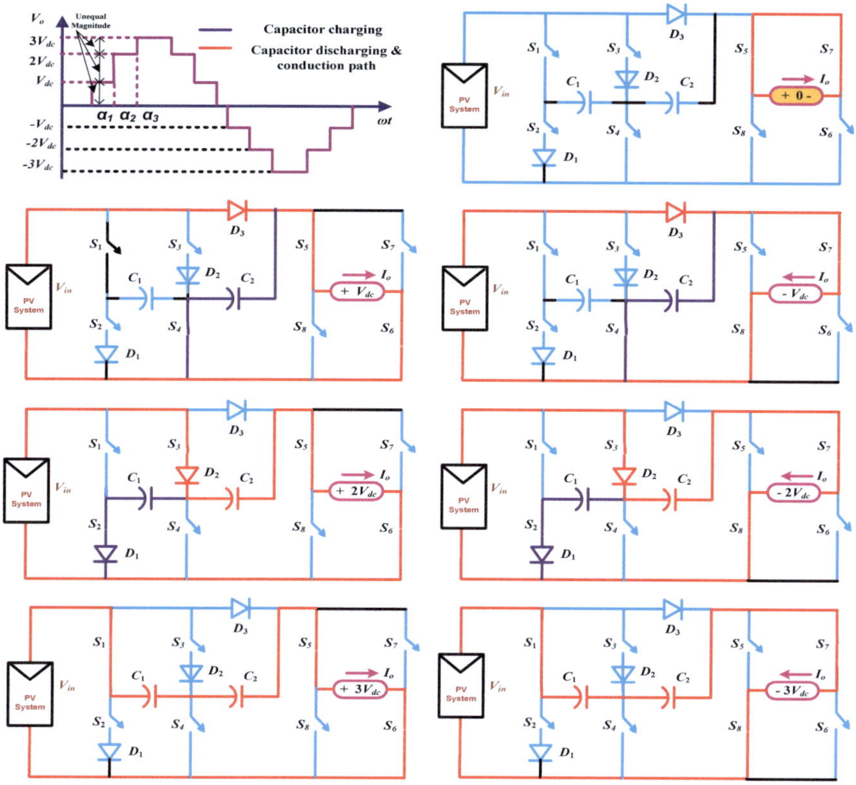

Fig. 8 Working operation of 7L SCMLI

of the voltage output waveform can be described in Fourier series as follows:

$$V(\omega t) = \sum_{n=1}^{\infty} V_n \sin(n\omega t) \tag{8}$$

Here, V_n is the magnitude of nth order of harmonic. Automatically get canceled even order of harmonics due to quarter wave balance, i.e., $b_n = 0$ for all n. Hence, the obtain new equation of output voltage becomes

$$V_n = \begin{cases} \dfrac{4V_{dc}}{n\pi} \displaystyle\sum_{i=1}^{\infty} K_i \cos(n\alpha_i), & \text{odd } (n) \\ 0, & \text{even } (n) \end{cases} \tag{9}$$

Here, V_{dc} is known as basic voltage and the ratio V_{dci} and V_{dc} known as K_i. For example, $V_{dc1} = V_{dc} \times K_1$.

In this work of paper by using APSO techniques targeted for reduce the low-order harmonics. The known voltage steps and switching angles, $B$is considering as unknown variables to solve the equation set.

$$M = \frac{1}{B}\,[K_1 \cos\alpha_1 + K_2 \cos\alpha_2 + K_3 \cos\alpha_3 + K_4 \cos\alpha_4 + \ldots.. + K_L \cos\alpha_L] \quad (10)$$

$$M \text{ is define as under} : M = \frac{V_1}{4BV_{dc}/\pi}(0, M < 1) \quad (11)$$

To obtain the less% THD, it is necessary to eliminate the dominant low-order harmonics from voltage output. Hence, the three nonlinear equations for the elucidation of the problem are suspicious in this work as under:

$$\begin{aligned}
&\frac{1}{4}[K_1(\cos\alpha_1) + K_2\cos(\alpha_2) + K_3\cos(\alpha_3)] = M \\
&K_1(\cos(3\alpha_1) + K_2\cos(3\alpha_2) + K_3\cos(3\alpha_3) = 0 \\
&K_1(\cos(5\alpha_1) + K_2\cos(5\alpha_2) + K_3\cos(5\alpha_3) = 0 \\
&K_1(\cos(7\alpha_1) + K_2\cos(7\alpha_2) + K_3\cos(7\alpha_3) = 0
\end{aligned} \quad (12)$$

For the definite modulation index, discontinuity occurs due to the context of these equations. Thus, the SHE problem nonlinear equations are resolved as an optimization problem. The above problem may be overcome using predictable techniques like the Newton–Raphson (NR) method, but these need harder computations and a lot of calculation time. Although they converge sluggish genetic algorithms are a type of random search processes used to identify estimated solutions of optimization problems. The objective function with the best value of switching angles and faster converges is solved using the APSO technique in this paper.

Voltage THD has been considered as an objective function in the work which is then got reduced by the elimination of selected harmonic order, third, fifth and seventh with dissimilar value of switching angle. Hence, the optimization process becomes as under:

$$\text{Minimize, } \%\text{THD} = \left[\frac{1}{V_1^2}\sum_{K=2}^{\infty}(V_K)^2\right]^{\frac{1}{2}} \times 100 \quad (13)$$

where V_k is represent the harmonics voltage.

$$\text{Subject to condition that, } 0 < \alpha_1 < \alpha_2 < \alpha_3 \le \frac{\pi}{2} \quad (14)$$

7 Anti-predatory Particle Swarm Optimization

Anti-predatory Particle Swarm Optimization (APSO), a novel PSO variation with anti-predatory qualities, was suggested by Selvakumar and Thanshkodi. The anti-predatory method's foundation is bird flocking, which helps the swarm or particle (bird) avoid its predators. APSO's working behavior is composed of two parts: (a) cognitive behavior, which is separated into two categories: horrible experience and good quality knowledge and (b) social behavior, which is separated into two categories: global worst experience and global good experience.

A swarm superior knowledge component reflects its recollection of the previously chosen best location, but horrible experience component refers to the worst spot it has visited. While food sites are represented using the locations of the best points in a typical PSO, predators are regarded as the least desirable points in the current work.

The new modify update position–velocity update equation in the APSO system is suitably adjusted by Eqs. (15) and (16):

$$\begin{aligned} V_{ij}^{r+1} &= w^r v_{ij}^r + Z_1R_1\left(Pb_{ij}^r - P_{ij}^r\right) + Z_2R_2\left(Pw_{ij}^r - P_{ij}^r\right) \\ &\quad + Z_3R_3\left(Gb_j^r - P_{ij}^r\right) + Z_4R_4\left(Gw_j^r - P_{ij}^r\right) \end{aligned} \tag{15}$$

$$P_{ij}^{r+1} = P_{ij}^r + V_{ij}^{r+1} \tag{16}$$

7.1 APSO Algorithm for THD Calculation

Step 1. Initiate the modulation index (MI) term, population size, number of maximum iteration (itermax) and coefficients of acceleration, i.e., Z_1, Z_2, Z_3, and Z_4 are initialized. $Z_1$and Z_2 Values lies within the range 0.1–2 and Z_3 and Z_4 are position to 2. Randomly makes a set of switching angles, *i.e.*, α1, α2 and α3.

Step 2. The values of switching angles should satisfy the practicability criteria to it lies in between [0, 90], i.e., $0 < \alpha_i < 90$ for i = 1, 2, 3.

Step 3. Using the formula, feasibility of the particle is checked maintain in below:

$$\gamma(x) = e^{\left(\sum_{i=1}^{m} log(1+g(x)) + \sum_{i=m}^{n} log(1+h(x))\right)} \tag{17}$$

If $\gamma(x)$=1, x is feasible and If $\gamma(x)$>1, x is infeasible.

Here, $g(x)$ and $h(x)$ are the dissimilarity and similar constraints, respectively.

Step 4. The fitness function *f(x)* for every set of switching angles is determined, if the

Step 5. Calculation value is feasible then as it is value is considered; otherwise, improvement factor are added in function. Correction factor are $|(C_1{*}N + C_2{*}S(x))|$, where $S(x)$ denotes the summary of all constraints; C_1 and C_2 are correction coefficients and N denotes the number of constraints.

Step 6. The starting positions of swarms until iteration r are measured as best local location of the particle, *i.e.*, Pb_{ij}^{r} and best global location Gb_{j}^{r} are the fitness function $f(x)$ of the minimum value. Correspondingly, the closing location of swarms are taken as worst local position, *i.e.*, Pw_{ij}^{r} and the location of global worst Gw_{j}^{r} are the highest value of the fitness function $f(x)$.

Step 7. The position and velocity for particle are modified by Eqs. (15) and (16). Finally, the best location of swarms is calculated by comparing the new location of swarm population with original location of particles in swarms and modifying the iteration count.

Step 8. When the iteration count attained the highest set value then determination the fitness function is calculated by changing the modulation index from 0.1 to 0.9

Step 9. Otherwise, do again Steps 4–7.

Working processes of APSO algorithms in flow chart pattern are shown in Fig. 9.

8 Comparison of GA with the Proposed APSO Algorithm

Upon evaluating the properties of the GA and APSO algorithms, it is discovered that both methods converge to generate feasible solutions at every MI. First, both methods converge at 0.9MI if the population size is 100. GA was able to determine the objective function of 13.9921% in 200 generations, whereas APSO converged to a more efficient solution of 11.8216% in 100 iterations. The result shows that the APSO algorithm converges faster than the GA algorithm.

Additionally, a study is conducted in a MATLAB/Software environment on an Intel(R) core (TM), i5-2540 M CPU at 2.60 GHz, logging computing time and achieving the objective value into kindness while keeping all other scenarios fixed in order to verify the approach's robustness. For every 100-person population, the GA and APSO algorithms are run 20 times at 0.5, 0.7, and 0.9 MI.

Table 4 represents the best objective value with consequent calculation times of 100, 150, and 300 iterations.

The comparison of Table 4 shows that under same conditions, APSO always converges faster than GA and also approaches to a better solution.

Figure 10a shows the variation of switching angles obtained using APSO for different M. The switching angles satisfy the condition, $0 < \alpha_1 < \alpha_2 < \alpha_3 < \frac{\pi}{2}$. The plot of THD obtained using both the algorithms (APSO and GA) against M is depicted in Fig. 10b taking into account a population size of 100 individuals. It is clearly observed that APSO has given lesser THD value at full range of M using APSO when compared to GA. This accomplishment also shows that APSO

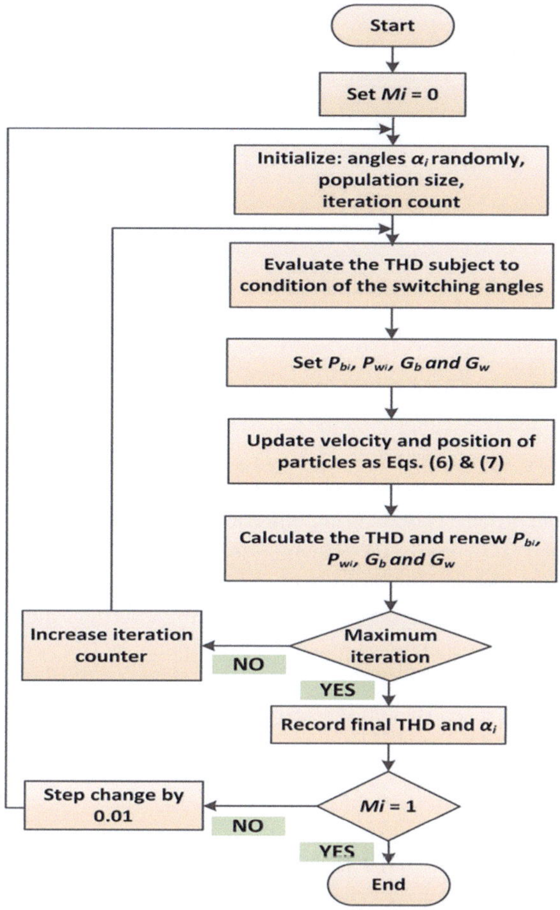

Fig. 9 The flowchart of the APSO algorithm

Table 4 GA and APSO comparison behavior for 100 individuals per population

MI	Iteration = 100		Iteration = 150		Iteration = 300	
	GA	APSO	GA	APSO	GA	APSO
Best objective value in 20 runs, in %						
0.5	18.8894	16.65	16.819	14.652	7.9712	5.6687
0.6	16.6681	14.263	12.27	8.942	6.4593	4.9987
0.9	12.9621	11.8216	8.131	5.9536	4.8919	4.0012
Computational time, in sec						
0.5	0.8877	0.6439	2.7543	2.0056	5.6542	3.9768
0.6	0.8797	0.4542	2.5512	1.9988	4.8976	3.7671
0.9	0.6523	0.3467	2.2091	1.5678	4.7531	2.5578

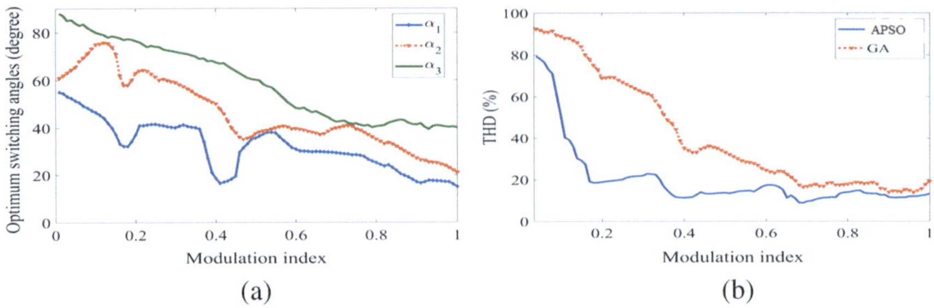

Fig. 10 **a** Switching angles obtained using APSO, **b** Comparison of THD (%) against M for GA & APSO

converges better than the GA. THD decreases with increase in *M*, which satisfies the operational principle of MLIs.

Additionally, in order to demonstrate the algorithm's resilience, a test is carried out in MATLAB on an Intel(R) core (TM), i7, 2.60 GHz processor, taking into account computing time and the achieved objective value under the assumption that all other parameters remain constant. Both algorithms carry out 20 times at 0.5, 0.6, and 0.9 modulation indices, using 100 individuals per population. The optimal objective value and associated computation times for 100, 150, and 300 iterations are displayed in Table 4. Table 4 comparison illustrates that, in the identical circumstances, APSO continuously converges more quickly than GA and moves closer to a superior solution.

9 Results and Simulation

The7L-PV-SCMLI topology proposed in this paper consists of less number of components comparatively traditional MLI for generating for 7L output voltage with additional features like less distortion, less EMI, self-balancing capability and minimal THD values. The proposed 7L-PV-SCMLI consisting of semiconductor IGBT switches and PV system with Fuzzy MPPT technique as shown in Fig. 8. This work is carried out using version 2021 at MATLAB/Simulink platform. For confirmation purpose, 80 V single dc source consider instead of PV system. The specifications of various components used in this system are as follows: Output frequency of the inverter is 50 Hz, capacitor value is given to be 2000μF, the values of both R and RL loads are 150 Ω and 180 Ω-500Mh and 100 Ω-180 mH, respectively. For computing the switching angles for corresponding switches of the proposed MLI, SHE-based APSO technique is used. Different MI values by using the computer programming calculated. The three different nonlinear equation solved regards to targeting to reduce the third-, fifth-, and seventh-order harmonics. Corresponding angles to 0.95 *MI* (find optimal values of angles) are used in this simulation work. At a low value of 0.2, *MI* is also taken in this work for verify performance of harmonic control for 7L-PV-SCMLI.

Figure 11a represents the PWM pulses from $gs1(V)$ to $gs8(V)$ and Fig. 11b shows the total standing voltage from $Vs1(V)$ to $Vs8(V)$.

Capacitor voltages ($V_{C1}\&V_{C2}$) and output voltage and current with R-load (from 0 to 0.1 s) and RL-load behavior (from 0.1 to 0.14 s) are depicted in Fig. 12a and b. The distortion is less in the output voltage and current between 0.1 and 0.14 s.

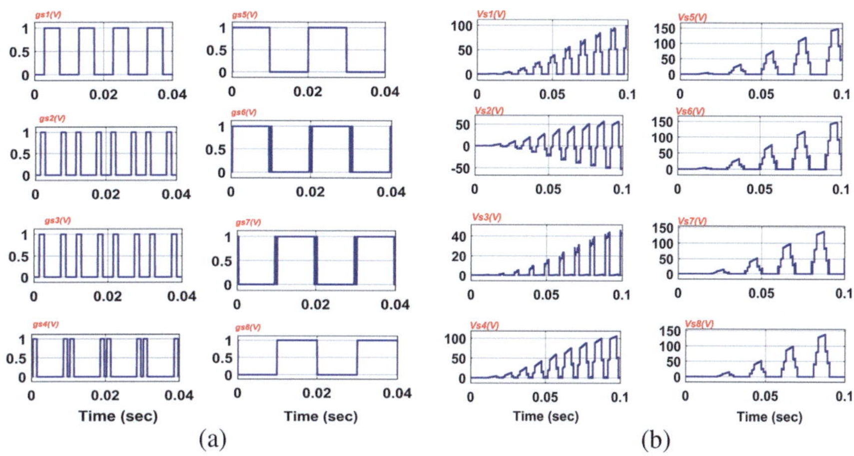

Fig. 11 **a** PWM pulses from $gs1(V)$ to $gs8(V)$ and **b** total standing voltage from $Vs1(V)$ to $Vs8(V)$

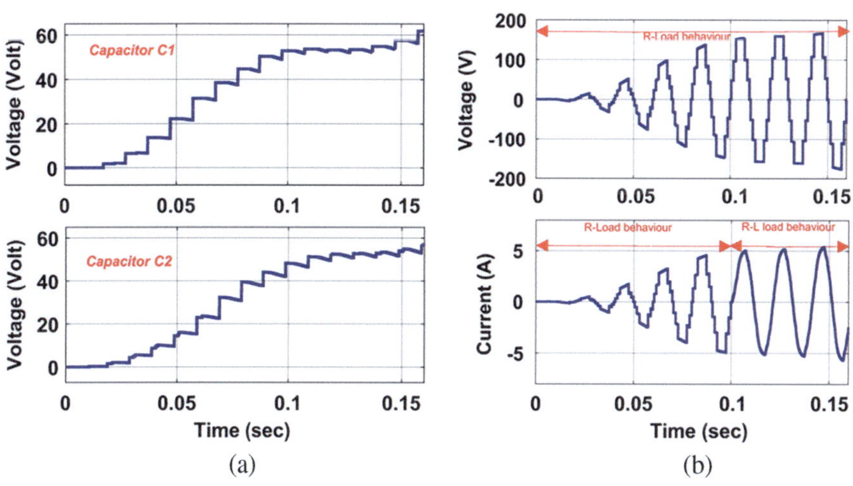

Fig. 12 Waveform of output voltage: **a** Capacitor voltage ($C_1\&C_2$), **b** 7L-PV-SCMLI voltage and current output

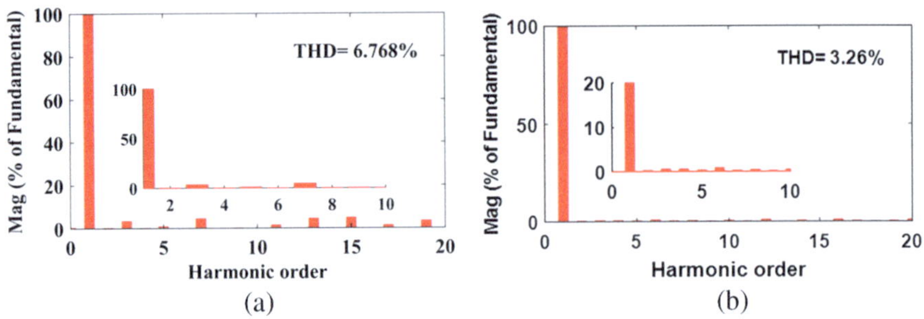

Fig. 13 FFT analysis of output voltage for different modulation index: **a** At MI value is 0.2 **b** MI value is 0.9

Figure 13a and b represents the voltage THD values of different *MI,* i.e., 0.2 and 0.9. At the 0.2*MI* %THD is 6.786% and 0.9*MI* %THD is 3.26%, it can be seen that MLI operates satisfactorily with different *MI* values and also satisfactorily reduced lower order harmonics (i.e., 3th, 5th, and 7th).

10 Conclusion

The switched capacitor architecture is used in this research study to design the 7L-PV-SCMLI setup and concept. The recommended 7L-PV-SCMLI architecture just requires one PV source and a few power semiconductor components. The optimal power from PV systems is produced by FLC MPPT techniques with the least amount of oscillation. Because capacitors are self-balancing, the suggested MLI structures can produce output voltage and current those are tripled. The construction of well-known MLIs is compared with this one based on the number of components, TSV, and voltage enhancement capacity that can support different loads. The switching angles of IGBTs are found using the new PSO version, or APSO technique, which uses fundamental frequency to justify very low loss operation. Under the same conditions, the GA and APSO are evaluated for a population of 100 individuals; APSO consistently converges faster than GA and approaches a better solution. For energy conversion applications needing low to medium power, the proposed MLI structure is perfect.

References

1. El Hammoumi A, Chtita S, Motahhir S, El Ghzizal A (2022) Solar PV energy: From material to use, and the most commonly used techniques to maximize the power output of PV systems: A focus on solar trackers and floating solar panels. Energy Rep 8:11992–12010. https://doi.org/10.1016/j.egyr.2022.09.054

2. Hernández-Callejo L, Gallardo-Saavedra S, Alonso-Gómez V (2019) A review of photovoltaic systems: Design, operation and maintenance. Sol Energy 188:426–440. https://doi.org/10.1016/j.solener.2019.06.017
3. Gopal Y, Panda KP, Birla D (2022) Genetic algorithm and anti-predatory swarm optimisation-based solutions for selective harmonic elimination in multilevel inverters. Int J Power Electron 15: 290–308. https://doi.org/10.1504/IJPELEC.2022.12240
4. Saiprakash C, Mohapatra A, Nayak B, Ghatak SR (2021) Analysis of partial shading effect on energy output of different solar PV array configurations. Mater Today: Proc 39 (Part 5):1905–1909. ISSN 2214–7853. https://doi.org/10.1016/j.matpr.2020.08.307
5. Youssef AR, Hefny MM, Ali AIM (2023) Investigation of single and multiple MPPT structures of solar PV-system under partial shading conditions considering direct duty-cycle controller. Sci Rep 13:19051. https://doi.org/10.1038/s41598-023-46165-1
6. Baba AO, Liu G, Chen X (2020) Classification and evaluation review of maximum power point tracking methods. Sustain Futur 2:100020. ISSN 2666–1888. https://doi.org/10.1016/j.sftr.2020.100020
7. NMCM, Al-Durra A, VK, JP, EL-Fouly THM, Zeineldin H A new multilevel inverter with reduced component count for a standalone solar energy conversion system. In: IEEE Transactions on Industry Applications. https://doi.org/10.1109/TIA.2023.3331672
8. Gopal Y, Panda KP, Kumari A, Rosas-Caro JC (2023) A switched-capacitor-based 7-level self-balancing high-gain inverter employing a single DC source. Int Trans Electr Energy Syst, Article ID 5545081. https://doi.org/10.1155/2023/5545081
9. Srivastava A, Seshadrinath J (2023) A single-phase seven-level triple boost inverter for grid-connected transformerless PV applications. IEEE Trans Industr Electron 70(9):9004–9015. https://doi.org/10.1109/TIE.2022.3215815
10. Gopal Y, Kumar YNV, Kumari A, Prakash O, Chowdhury S, Almehizia AA (2023) Reduced device count for self balancing switched-capacitor multilevel inverter integration with renewable energy source. Sustainability 15:8000. https://doi.org/10.3390/su15108000
11. Babu L, Chacko M (2024) Modular multilevel converter with a Sensing and Balancing (SAB) technique for capacitor voltage balancing. Iran J Sci Technol Trans Electr Eng 48:251–264. https://doi.org/10.1007/s40998-023-00659
12. Jena K, Gupta KK, Kumar D et al (2024) Single-stage single-phase 5-level switched-capacitor multilevel inverter. Iran J Sci Technol Trans Electr Eng 48:265–276. https://doi.org/10.1007/s40998 023 00660 3
13. Choi J, Kang F (2015) Seven-level PWM inverter employing series-connected capacitors paralleled to a single DC voltage source. IEEE Trans Industr Electron 62:3448–3459
14. Gopal Y, Birla D, Lalwani M (2020) Reduced switches multilevel inverter integration with boost converters in photovoltaic system. SN Appl. Sci. 2:58. https://doi.org/10.1007/s42452-019-1848-7
15. Kumari A, Gopal Y, Dhaked DK, Panda KP, Kumar YV (2023) A single source five-level switched-capacitor based multilevel inverter with reduced device count. e-Prime—Advances Electr Eng, Electron Energy 5:100235. https://doi.org/10.1016/j.prime.2023.100235
16. Dubey S, NarotamSarvaiya J, Seshadri B (2013) Temperature Dependent Photovoltaic (PV) efficiency and its effect on PV production in the world—a review. Energy Procedia 33:311–321. https://doi.org/10.1016/j.egypro.2013.05.072
17. Mohapatra A, Nayak B, Das P, Mohanty KB (2017) A review on MPPT techniques of PV system under partial shading condition. Renew Sustain Energy Rev 80:854–867. https://doi.org/10.1016/j.rser.2017.05.083
18. Bouselham L, Hajji M, Hajji B, Bouali H (2017) A new MPPT-based ANN for photovoltaic system under partial shading conditions. Energy Procedia, 111:924–933. https://doi.org/10.1016/j.egypro.2017.03.255
19. Ye Y, Cheng KEW, Liu J, Ding K (2014) A step-up switched- capacitor multilevel inverter with self-voltage balancing. IEEE Trans Ind Electron 61:6672–6680
20. Li X, Wen H, Hu Y, Jiang L (2019) A novel beta parameter based fuzzy-logic controller for photovoltaic MPPT application. Renew Energy 130:416–427. https://doi.org/10.1016/j.renene.2018.06.071

21. Rezk H, Aly M, Al-Dhaifallah M, Shoyama M (2019) Design and hardware implementation of new adaptive fuzzy logic-based MPPT control method for photovoltaic applications. IEEE Access 7:106427–106438. https://doi.org/10.1109/ACCESS.2019.2932694
22. López et al (2020) Carrier-based PWM equivalent to multilevel multiphase space vector PWM techniques. In: IEEE Transactions on industrial electronics, vol 67, no 7, pp 5220–5231. https://doi.org/10.1109/TIE.2019.2934029

Common-Ground Boosting Multilevel Inverter Topology with Optimal Components Count

Kaibalya Prasad Panda, Sumant Kumar Dalai, Yatindra Gopal, and Gayadhar Panda

Abstract The behavior of output voltage and leakage current in solar photovoltaic (PV) applications are significant concerns in transformerless grid-connected multilevel inverters (MLIs). A novel Boosting Common-Ground type MLI (BCGMLI) with an optimal component count is demonstrated in this paper. It provides a step-up output voltage while using less components than most state-of-the-art topologies. Improved voltage control, reduced component count, self-balanced capacitor-based boosting, and higher efficiency are all benefits of the inverter topology, which is developed to address significant issues in PV systems. In order to generate a 9-level two-fold phase voltage, BCGMLI uses just eight complementary operational switches. The proposed inverter architecture successfully resolves the problem of leakage current through its common grounding feature. In addition, this new inverter is capable of operating under sudden changes in modulation index, load disturbance conditions, sudden changes in input voltage magnitude, and managing reactive power when the grid experiences lagging or leading power factor conditions. With component design guidelines, in-depth explanation of the inverter's circuit operation, control strategy, pulse width modulation (PWM) method, and loss analysis is presented. The effectiveness and feasibility of the BCGMLI is demonstrated by showcasing experimental results.

K. P. Panda (✉)
Department of Electrical Engineering, School of Energy Technology, Pandit Deendayal Energy University, Gujarat, India
e-mail: kaibalyapanda.nit@gmail.com

S. K. Dalai
National Institute of Technology, Meghalaya, India

Y. Gopal
Department of Electrical Engineering, National Institute of Technology Arunachal Pradesh, Jote, Arunachal Pradesh, India

G. Panda
National Institute of Technical Teachers Training and Research, Kolkata, India

Y. Gopal et al. (eds.), *Multilevel Inverter Technology for Renewable Energy Systems*, Green Energy and Technology, https://doi.org/10.1007/978-3-032-03589-9_7

Keywords Common ground · Leakage current · Nine-level inverter · Switched-capacitor cell · Boosting common-ground multilevel inverter (BCGMLI) · Transformerless inverter

1 Introduction

Multilevel inverters (MLIs) have become focal to the conversation of power electronics in recent years due to its diverse applications in many fields. MLIs may utilise several dc sources, diodes, capacitors, and switches connected to each other which are turned on and off to produce stepped output which is closer to a sinusoidal wave [1]. This enables MLIs to have many attractive features over their counterparts, such as better power quality due to reduced harmonics, lower voltage stress and improved efficiency. Therefore, MLIs find use case in renewable power generation, electric vehicles, high-frequency ac power distribution, power quality filters etc. [2]. Cascaded H-Bridge type MLI (CHB MLI) is highly modular, making it easy to increase the number of voltage levels in the output. But this comes with the major drawback, which is the requirement of multiple isolated dc sources and a very high switch count. On the other hand, Neutral Point Clamped MLI (NPC MLI), and Flying Capacitor (FC MLI) type of topologies suffer from the voltage balancing issues and the total components count per number of voltage levels.

In recent literature, there has been a focus on developing single-input MLIs to streamline the circuit complexity. One approach discussed in [3] involves utilizing a single input MLI comprising series-connected capacitors to generate a 7-level output. However, this method lacks voltage boosting capability, necessitating additional voltage balancing control for maintaining the required voltage across capacitors. Authors in [4] proposed a single-input structure that facilitates self-voltage balancing through the incorporation of additional switches. These switches enable parallel charging of capacitors and series discharge with the input, thus achieving the desired voltage levels. The concept of series–parallel charge balancing is elaborated in [5], introducing a single-input switched-capacitor MLI (SCMLI). Furthermore, [6] introduces a self-balanced SCMLI utilizing two H-bridge units, resolving complexity issues associated with driver circuit requirements. While various advancements have been made to alleviate voltage stress in SCMLIs, recent studies such as [7] explore single-phase and three-phase extendibility without requiring a full-bridge. However, these approaches often entail trade-offs, such as increased switch count or capacitor requirement.

For an inverter to be used in the solar PV grid integration system, the inherent voltage boosting feature is essential to avoid the additional boosting circuit. One of the generalized boosting MLI introduced in [8] demands a substantial number of switches in symmetrical and asymmetrical extension. On the other hand, the single-dc MLI described in [9] utilizes series diodes but falls short in operability under low power factor (PF) loading conditions. Another design, as outlined in [10], presents a 9-level single input MLI employing two capacitors and two diodes in its front-end circuit.

Despite both capacitors being equally charged to half of the input voltage magnitude, the voltage boosting is only twice the input. Furthermore, the incorporation of an H-bridge at the back-end in these configurations increases the TVS. To address this concern, alternative generalized MLIs have been proposed in [11] and [12] without employing a back-end full bridge. However, achieving higher voltage levels at the output necessitates multiple symmetrical and asymmetrical sources. Although attempts have been made in [13] to reduce the number of capacitors in the circuit topology, it remains unable to eliminate the requirement for multiple sources. In efforts to mitigate this challenge, 9-level MLIs have been developed in [14, 15] utilizing a single dc source and minimizing the number of capacitors. Topologies proposed in [16] and [17] are quadruple boosted but have a much higher TSV than the previously mentioned topologies.

One of the major problems associated with all these inverter topologies is that they do not utilize a transformer for grid side connection. This leads to a rise of isolation issues and a high leakage current which arises due to the parasitic capacitance between the high-voltage output terminals and the ground. The capacitance allows a small amount of AC current to flow to ground, resulting in leakage current. This phenomenon becomes particularly pronounced in multilevel inverters due to their high switching frequencies and the presence of multiple voltage levels. MLI in [18] utilizes a single dc source with reduced component count to provide double boosted output at a lower TSV. Another lucrative solution for complete elimination of leakage current from MLI topologies is the utilization of common ground-based structures. The common ground is yielded by providing a short-circuited path through a grounded potential between the negative terminal of the input dc source and the neutral point of the grid. A common ground topology in [19] utilizes eight switches, two diodes, and two capacitors along with two dc sources to provide a nine-level stepped output. The two dc sources are an undesirable feature and topology proposed in [20] overcomes this challenge. It utilizes fewer components but has much higher peak inverse voltage than [21]. Literature [22] proposes a topology that provide continuous dc current while using a single source, nine switches, two diodes and four capacitors. Therefore, a new and improved Boosting Common-Ground type MLI (BCGMLI) topology which tries to overcome the challenges mentioned in the above literature is proposed in this work.

2 Proposed Boosting 9L Inverter Topology

A. Circuit Description

The proposed 9-level inverter structure is schematically shown in Fig. 1. The inverter design has nine power switches (S_1–S_9), two diodes (D_1 and D_2), three switching capacitors (C_1, C_2, and C_3), and a single DC source (solar PV). A parasitic capacitance link between the grid and PV source makes common-mode voltage difference a problem. The IEEE leakage current standards require this discrepancy to be minimized for safety and regulatory compliance. The negative point of the load/grid neutral is strategically connected to the negative terminal of the DC source in the proposed topology. This connection short-circuits the parasitic path to nullify common-mode voltage.

The capacitors (C_1–C_3) maintain self-voltage balance with their reference values V_{in}, V_{in}, and $0.5V_{in}$, respectively. The suggested Pulse Width Modulation (PWM) method controls capacitor charging and discharging. Table I shows the switching states and the switched capacitors' charging and discharging events at specific output voltage levels. It is important to notice that this inverter has 9L, doubling input voltage. The output voltage has $\pm 0.5V_{in}$ increments, ranging from $+2V_{in}$ to $-2V_{in}$.

2 Operational Modes of the Proposed 9-Level Inverter

The suggested nine-level common ground grid-tied inverter's operational modes are discussed below. Figure 2a–j shows nine-level Multilevel Inverter (MLI) designs for different output voltage tiers. Table I also lists the switching states for each voltage level. Table 1 shows that charging and discharging SCs generate nine levels from the proposed topology. Different voltage-generating modes throughout the positive and negative half cycles are explained.

(a) $\pm$ **$2V_{in}$**: Conduction of switches S_1, S_5, S_6, and S_9 generates a peak output voltage of $+2V_{in}$ ($V_{in} + V_{C1}$), discharging SC C_1. The diode D_2 charges the

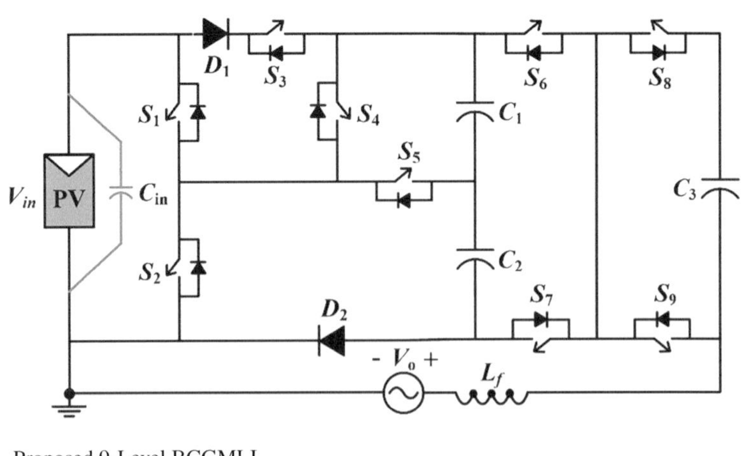

Fig. 1 Proposed 9-Level BCGMLI

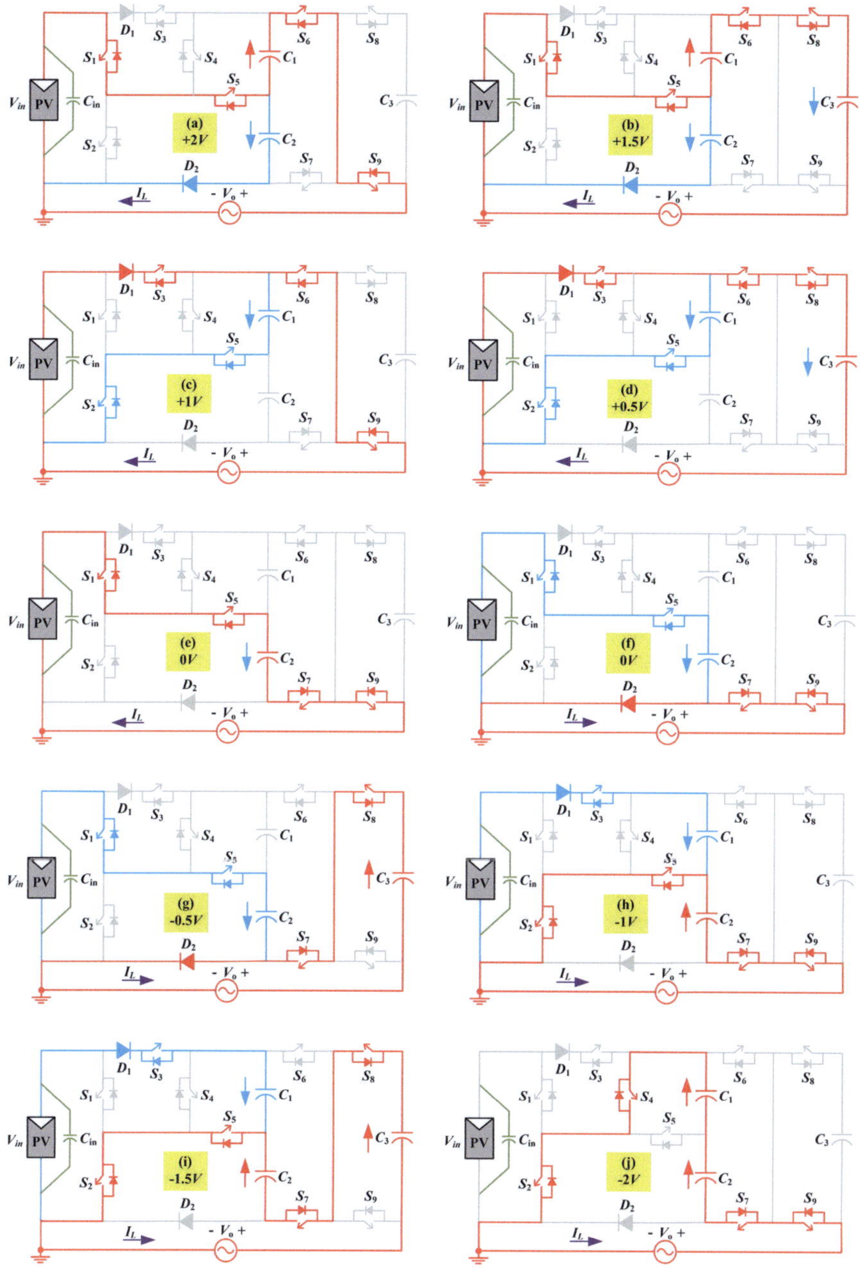

Fig. 2 Operating states and conduction path

Table 1 Switching states and their effect on sc's voltage state

State	Switching states									Diode states		Capacitor states			Output voltages
	S_1	S_2	S_3	S_4	S_5	S_6	S_7	S_8	S_9	D_1	D_2	C_1	C_2	C_3	V_O
a	1	0	0	0	1	0	1	0	1	0	0	–	↑	–	$+2V_{in}$
b	0	1	1	0	1	1	0	1	0	1	0	↑	–	↑	$+1.5V_{in}$
c	0	1	1	0	1	1	0	0	1	1	0	↑	–	–	$+1V_{in}$
d	1	0	0	0	1	1	0	1	0	0	1	↓	↑	↑	$+0.5V_{in}$
e	1	0	0	0	1	1	0	0	1	0	1	↓	↑	–	$0V_{in}$
f	1	0	0	0	1	0	1	0	1	0	1	–	↑	–	$0V_{in}$
g	1	0	0	0	1	0	1	1	0	0	1	–	↑	↓	$-0.5V_{in}$
h	0	1	1	0	1	0	1	0	1	1	0	↑	↓	–	$-1V_{in}$
i	0	1	1	0	1	0	1	1	0	1	0	↑	↓	↓	$-1.5V_{in}$
j	0	1	0	1	0	0	1	0	1	0	0	↓	↓	–	$-2V_{in}$

SC C_2 simultaneously. During the negative half cycle, switches S_2, S_4, S_7, and $S_9$are on. The discharged SCs C_1 and C_2produce an output voltage of $-2Vin$ ($-V_{C1} - V_{C2}$). While both half cycles generate levels, SC C_3 is idle. The conduction path is given in Fig. 2a and j.

(b) **$\pm$ 1.5V_{in}**: Switches S_1, S_5, S_6, and S_8 generate the third positive voltage level. SC C_1 discharges, charging SC C_3, yielding $+ 1.5V_{in}$ ($V_{in} + V_{C1} - V_{C3}$). The diode D_2 charges the SC C_2 simultaneously. S_2, S_3, S_5, S_7, and S_8 are ON during the negative half cycle, connecting the PV panel to the grid. The SCs C_2 and C_3 discharge to provide the third negative output voltage level, -1.5Vin ($-V_{in}$ $-$ $0.5V_{in}$). The SC C_1 charges through switches S_2, S_3, and S_5 and diode D_1 simultaneously. Figure 2b and i illustrate the load current direction for both half cycles as this output voltage level is generated.

(c) **$\pm$ 1V_{in}**: The switches S_2, S_3, S_5, and S_9 generate the second positive and negative voltage levels. During the positive half cycle, the PV panel and grid are linked via diode D_1 and switches S_3, S_6, and S_9, while SC C_1 is charged. The capacitors C_2 and C_3 are inactive. Negative voltage is created by discharging SC C_2 through switches S_7 and S_9, resulting in $-V_{in}$ ($-V_{C2}$). The SC C_1 is charged simultaneously by diode D_1 and switches S_2, S_3, and S_5. The SC C_3 is inactive. The conduction path is given in Fig. 2c and h.

(d) **$\pm$ 0.5V_{in}**: The first positive voltage level is created by conducting switches S_2, S_3, S_5, S_6, and S_8. SC C_3 is charged while SC C_1 is charged via switches S_2 and S_5. This produces an output voltage of $+ 0.5V_{in}$ ($+0.5V_{in}$ $-$ $0.5V_{in}$). The SC C_2 remains. During the negative half cycle, switches S_1, S_5, S_7, and S_8 are activated, and I_L moves from neutral to grid. An output voltage of $-0.5V_{in}$ ($-V_{C3}$) is produced. The SC C_1 is idle. Figure 2d and g show the load current direction for both half cycles to get this output voltage level.

(e) **$0V_{in}$**: In this mode the switches S_1, S_5, S_7 and S_9 are ON for both the half cycles connecting the PV panel and the grid. In the positive half cycle, SC C_2 simultaneously charges and the output voltage level generated is 0V ($V_{in} - V_{C2}$). While in the negative half cycle, the SC C_2 undergoes charging through switches S_1, S_5 and diode D_2 which produces a zero-level output voltage, 0V ($V_{in} - V_{C2}$). The SC C_1 and C_3 remains idle and the conduction path is as shown in Fig. 2e and f for both cycles.

3 Design Of Switched Capacitors

In the proposed 9L MLI, capacitors charge and discharge at switch table-defined intervals. Capacitors in parallel with the source charge to the input voltage. To increase output, the MLI switches are turned on and off to link this capacitor in series. The capacitor in series with the output load causes a voltage drop while discharging, resulting in power losses. The capacitor switches between charging and discharging, causing modest but frequent voltage decreases. When the capacitor is continuously linked in series with the output, voltage drops increase, resulting in higher power losses. This particularly generates voltage fluctuations between these capacitors. Capacitance and discharging time determine a capacitor's voltage range. Capacitance is chosen to produce different output voltages with acceptable ripple. Charging/discharging time and maximum voltage ripple determine capacitance. To calculate capacitance values, waveforms of V_O, V_{C1}, and V_{C3} are analyzed and deductions made. LDT can be calculated using the fundamental time period (T) at fundamental frequency ($\omega = \frac{2\pi}{T}$) and modulation index (1), with time instants are given as [23, 24].

$$t_1 = \frac{1}{\omega}sin^{-1}\left(\frac{1}{4m}\right), t_2 = \frac{1}{\omega}sin^{-1}\left(\frac{1}{2m}\right), t_3 = \frac{1}{\omega}sin^{-1}\left(\frac{3}{4m}\right) \quad (1)$$

For an inductive load with peak load current (I_O), and impedance angle (ϕ), the instantaneous load current (i_O) given by (2)

$$i_O = I_O sin(\omega_t - \phi) \quad (2)$$

Charge drawn from C_1 is Q_{C1} is given by deducing the modes during which it discharges from Fig. 3. Evidently the discharging occurs in the + 2, + 1.5 and −2 cycle. Therefore, the longest discharging cycle is the one from t_2 to t_5. Here $t_5 = \frac{T}{2} - t_2$. Accordingly, the mathematical output for the charge drawn is given by (3).

$$Q_{C1} = \int_{t_2}^{t_5} i_O dt = \frac{I_O}{\omega}cos\emptyset\sqrt{4 - \frac{1}{m^2}} \quad (3)$$

The optimal capacitance is given by (4)

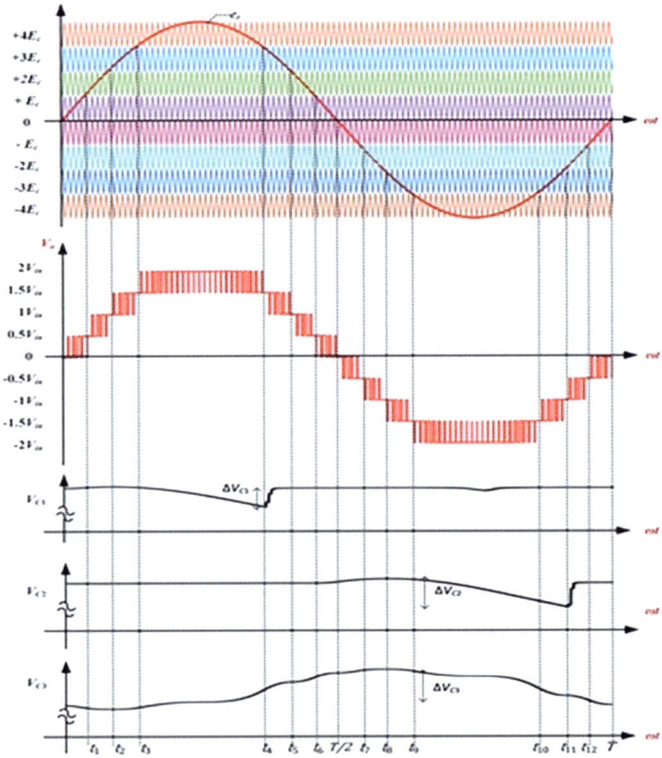

Fig. 3 PWM signal generation and SC status

$$C_1 = \frac{Q_{C1}}{\Delta V_{C1} \times V_{C1}} \tag{4}$$

Similarly, capacitors C_2, C_3 design also given by (5), (6);

$$C_2 = \frac{Q_{C1}}{\Delta V_{C2} \times V_{C2}} \tag{5}$$

$$C_3 = \frac{Q_{C3}}{\Delta V_{C3} \times V_{C3}} \tag{6}$$

3 Simulation and Experimental Results

The efficiency of the proposed layout has been proven by experimental analysis and simulation. The simulation is executed using the MATLAB/Simulink platform. The following parameters are used to perform the simulation: An R-L load with R = 50 Ω and L = 100 mH and switched capacitors with capacitances of 3.3 mF, 4.8 mF, and 4.9 mF, as well as a modulation index of 0.9, an output fundamental frequency of 50 Hz, and a carrier signal frequency of 5 kHz are all part of the setup.

Figure 4a shows the simulated inverter voltage, grid voltage, and currents from the simulation results for 200 V dc input. The inverter voltages V_O exhibit nine levels, while the grid voltage V_{GA} and current I_{GA} are sinusoidal. The results of the simulation under dynamic test conditions are displayed in Fig. 4b. Altering MI from 1, 0.75, and 0.6 lowers the voltage levels and lowers the capacitor voltage as well. The sudden drop in the input voltage from 200 to 160 V is shown in Fig. 4c. As long as the inverter voltage and capacitor voltage stay the same, the lagging to leading power factor is shown in Fig. 4d. As seen in Fig. 4e, there is a sudden step change from 10 to 15 A in the grid current.

Furthermore, experimental research are conducted under steady-state and dynamic test conditions. Figure 5a illustrates inverter voltage, grid voltage, and grid currents along with the capacitor voltages for 50 Hz operation, 110 V (RMS) grid voltage and 100 V dc input voltage. Figure 5b and c indicate that the grid voltage and current change under unity, lagging, and leading conditions, but that inverter and capacitor voltage had no effect on keeping it constant. Figure 5c also shows the grid current and capacitor voltage smoothly changes with change in power factor. By applying the charge-balancing concept, we can ensure that the voltages across the capacitors V_{C1}, V_{C2}, and V_{C3} are self-balanced at the target level, which in turn confirms the capacitors' two-time voltage boosting capability and their rate of stabilization at the target voltage. Most power losses come via capacitor charging-discharging and diode conduction, with the rest from switch action and other passive components. The proposed structure is appropriate for medium-voltage solar PV applications due to its high efficiency of 96.98% at 1 kW power rating.

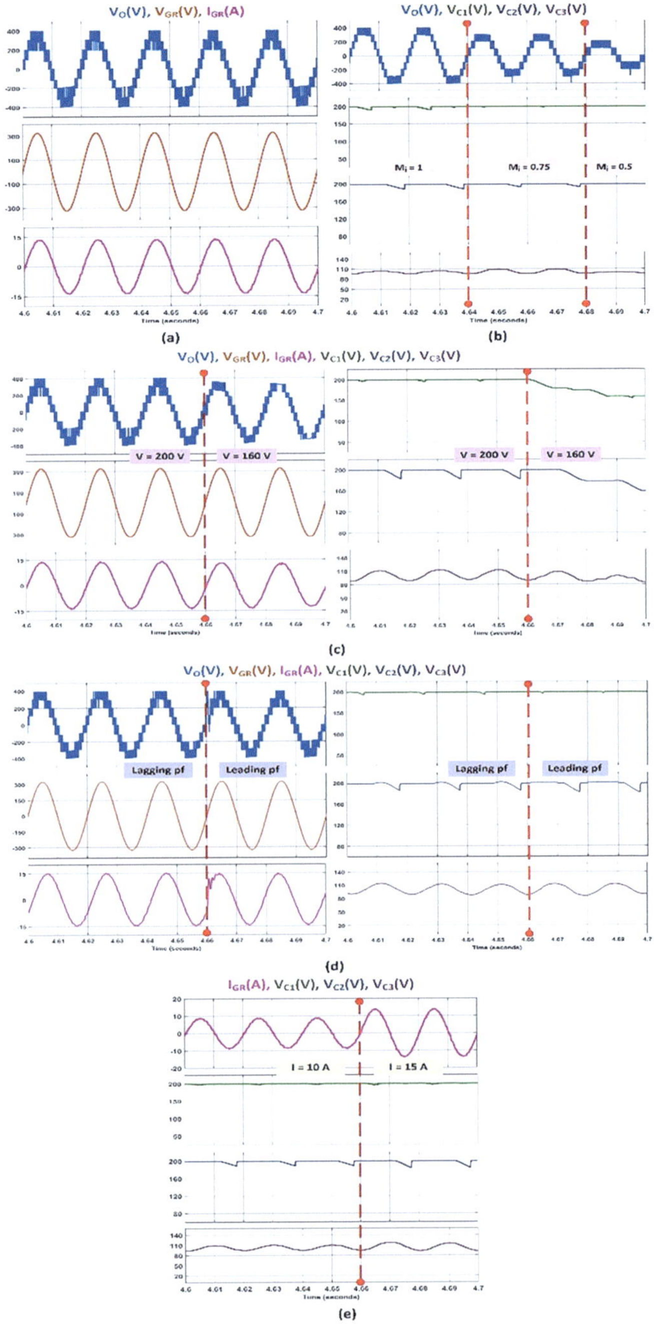

Fig. 4 Simulation results of inverter voltage, grid voltage, and current under **a** steady state, **b** variation in MI, **c** variation in input, **d** lagging to leading power factor, **e** step change of grid current respectively

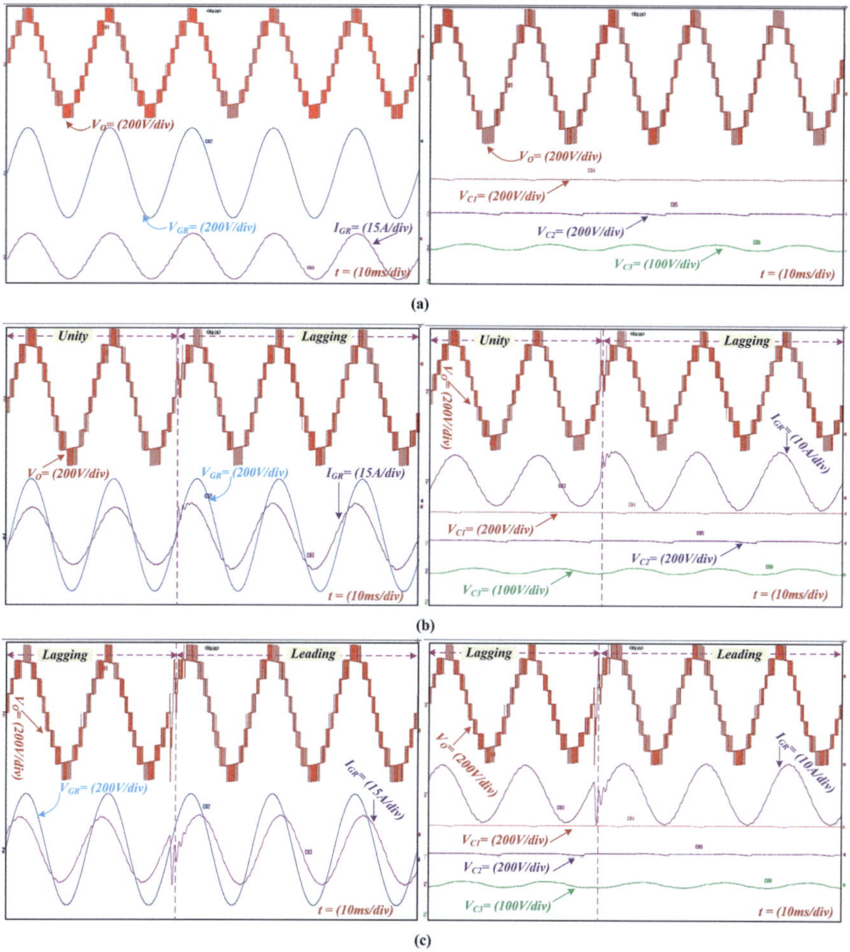

Fig. 5 Experimental results in grid-connected mode **a** inverter voltage, grid voltage, current, and capacitor voltage, **b–c** response in power factor from unity, lagging, and leading

4 Conclusion

In this paper, a new Common-Ground Boosting Multilevel Inverter Topology with Optimal Components Count was presented, offering significant improvements over conventional multilevel inverters. By reducing the number of components and integrating a boosting mechanism within the inverter itself, the proposed topology simplifies system design, improves power density, and reduces overall cost. The common-ground configuration addresses safety concerns and eliminates leakage currents, making the topology ideal for grounded PV systems and battery energy storage applications.

Furthermore, the optimized switching strategy ensures smooth voltage transitions, minimal total harmonic distortion (THD), and enhanced overall efficiency. Simulation and experimental results under different test conditions confirm the superior performance of this design, demonstrating its potential for scalability in various renewable energy and electric vehicle systems. The proposed inverter topology represents a promising advancement in the field of power electronics, providing a cost-effective, efficient, and practical solution for future energy applications.

References

1. Fong YC, Raman SR, Ye Y, Cheng KWE (2020) Generalized topology of a hybrid switched-capacitor multilevel inverter for high-frequency AC power distribution. IEEE J Emerg Sel Top Power Electron 8(3):2886–2897
2. Bana PR, Panda KP, Naayagi RT, Siano P, Panda G (2019) Recently developed reduced switch multilevel inverter for renewable energy integration and drives application: topologies, comprehensive analysis and comparative evaluation. IEEE Access 7:54888–54909
3. Barzegarkhoo R, Farhangi M, Lee SS, Aguilera RP, Siwakoti YP (2022) Nine-level nine-switch common-ground switched-capacitor inverter suitable for high-frequency AC-microgrid applications. IEEE Trans Power Electron 37(5):6132–6143
4. Gopal Y, Panda KP, Gupta PP, Dhaked DK, Arya Y (2025) Development and performance assessment of switched-capacitor multilevel inverters for solar PV applications. Sustain Energy Technol Assess 79:104360. ISSN 2213–1388
5. Lee SS, Lim CS, Lee K-B (2020) Novel active-neutral-point-clamped inverters with improved voltage-boosting capability. IEEE Trans Power Electron 35(6):5978–5986
6. Jakhar A, Sandeep N, Verma AK (2024) Common-ground-type quadruple boosting nine-level inverter. IEEE J Emerg Sel Top Ind Electron 1–8
7. Dhara S, Somasekhar V (2021) A nine-level transformerless boost inverter with leakage current reduction and fractional direct power transfer capability for PV applications. IEEE J Emerg Sel Top Power Electron
8. Babaei E, Gowgani SS (2014) Hybrid multilevel inverter using switched capacitor units. IEEE Trans Ind Electron 61(9):4614–4621
9. Ye Y, Cheng KWE, Liu J, Ding K (2014) A step-up switched-capacitor multilevel inverter with self-voltage balancing. IEEE Trans Ind Electron 61(12):6672–6680
10. Gopal Y, Kumari A, Panda KP, Dhaked DK, Arya Y (2025) Implementation and analysis of switched-capacitor multilevel inverters in solar photovoltaic system. Sustain Energy Technol Assess 75:104227. ISSN 2213-1388. https://doi.org/10.1016/j.seta.2025.104227
11. Zamiri E, Vosoughi N, Hosseini SH, Barzegarkhoo R, Sabahi M (2016) A new cascaded switched-capacitor multilevel inverter based on improved series-parallel conversion withless number of components. IEEE Trans Ind Electron 63(6):3582–3594. https://doi.org/10.1109/TIE.2016.2529563
12. Nakagawa Y, Koizumi H (2019) A boost-type nine-level switched capacitor inverter. IEEE Trans Power Electron 34(7):6522–6532
13. Kumari S, Sandeep N (2023) A self-balancing switched-capacitor-based dual boosting nine-level inverter. IEEE J Emerg Sel Top Power Electron 11(5):5311–5320
14. Panda KP, Kumar Dalai S, Panda G (2023) Step-up switched-capacitor common-grounded seven-level inverter for transformerless PV system. In: 2023 IEEE International conference on power electronics, smart grid, and renewable energy (PESGRE), Trivandrum, India
15. Mohamed Ali JS, Krishnasamy V (2019) Compact switched capacitor multilevel inverter (CSCMLI) with self-voltage balancing and boosting ability. IEEE Trans Power Electron 34(5):4009–4013

16. Siddique MD, Mekhilef S, Padmanaban S, Memon MA, Kumar C (2021) Single-phase step-up switched-capacitor-based multilevel inverter topology with SHEPWM. IEEE Trans Ind Appl 57(3):3107–3119
17. Sandeep N, Ali JSM, Yaragatti UR, Vijayakumar K (2019) Switched-capacitor-based quadruple-boost nine-level inverter. IEEE Trans Power Electron 34(8):7147–7150
18. Liu J, Lin W, Wu J, Zeng J (2019) A novel nine-level quadruple boost inverter with inductive-load ability. IEEE Trans Power Electron 34(5):4014–4018
19. Dalai SK, Panda KP, Siwakoti YP, Panda G (2024) Three-phase neutral-point-clamped multilevel inverter interfacing grid interactive renewable energy system. Int J Circuit Theory Appl. https://doi.org/10.1002/cta.4299
20. Saeedian M, Adabi ME, Hosseini SM, Adabi J, Pouresmaeil E (2019) A novel step-up single source multilevel inverter: topology, operating principle, and modulation. IEEE Trans Power Electron 34(4):3269–3282
21. Samizadeh M et al (2020) A new topology of switched-capacitor multilevel inverter with eliminating leakage current. IEEE Access 8:76951–76965
22. Yao J, Zhang Z, Zheng X, Ioinovici A (2024) A common ground switched capacitor multilevel inverter (SCMLI) with a low components count. IEEE Trans Aerosp Electron Syst 1–12
23. Dalai SK, Panda KP, Siwakoti YP, Panda G (2024) Three-phase switched-capacitor boost self-balanced multilevel inverter for photovoltaic applications. IEEE Trans Energy Convers 39(3):1818–1827
24. Srivastava A et al (2023) A new nine-level highly efficient boost inverter for transformerless grid-connected PV application. IEEE J Emerg Sel Top Power Electron 11(3):2730–2741

Investigation of Fault-Tolerant Multilevel Inverter Topologies

Dhananjay Kumar, Santosh Kumar Gupta, Vishal Rathore, Kasinath Jena, Ramjee Prasad Gupta, Samar Anand, Kumar Rahul, and Shahzad Ahsan

Abstract This study introduces various recently suggested multilevel power converters to improve the topology's dependability used in safety–critical applications. The proposed topologies improve the fault-tolerant (FT) capability in semiconductor power devices under any open-circuit (OC) fault. These topologies have generally reduced device MLIs (RD-MLIs). Furthermore, the proposed single-phase fault-tolerant (SPFT) topologies of the proposed circuit naturally achieve self-voltage balancing (SVB) ability, high efficiency and control effortlessness. Multi-carrier pulse-width modulation (PWM) technique is accepted to generate the switching pulses. Finally, the operating principles of the suggested FT-MLI topologies in a MATLAB/SIMULINK are verified by the experimental output obtained in the lab prototype.

Keywords Multilevel inverter · Reduce device (RD) · Reliability · Open circuit fault (OCF) · Pulse width modulation

D. Kumar (✉) · S. K. Gupta · S. Anand
Department of Electrical Engineering, Govternment Engineering College, Siwan, Bihar, India
e-mail: dhananjaymanit23@gmail.com

V. Rathore
Department of Electrical Engineering, Maulana Azad National Institute of Technology, Bhopal, Madhya Pradesh, India

K. Jena
Department of Electrical and Electronic Engineering, ARKA JAIN University, Jamshedpur, India

R. P. Gupta
Department of Electrical Engineering, MIT Muzaffarpur, Bihar, India

K. Rahul · S. Ahsan
Department of Electrical & Electronics Engineering, LNJPIT, Chhapra, Saran, India

Y. Gopal et al. (eds.), *Multilevel Inverter Technology for Renewable Energy Systems*, Green Energy and Technology, https://doi.org/10.1007/978-3-032-03589-9_8

1 Introduction

Due to increased power density and excellent efficiency, the power electronic converters are becoming more and more significant in a variety of applications like adjustable speed drives, electric vehicles, and utility interfacing of renewable energy resources (RES) with the grid, HVDC transmission system [1]. Large filter sizes and significant switching stresses are issues with traditional two-level converter-based topologies at higher power ratings. Multilevel inverters have occurred as a major result to these problems, offering extremely power-efficient operation and component scale reduction [2].

Researchers have studied a range of MLI topologies over the last few decades, such as cascaded H-bridge (CHB), flying capacitor (FC), and neutral-point clamped (NPC) topologies [3]. The CHB and NPC converters are two of the most used topologies in industrial settings. Among these MLI topologies, the CHB has attained high voltage and high-power levels. The MLI's high degree of scalability, which enables the inverter to reach high voltage using medium-voltage semiconductor power and switches, is its most significant differentiator [4]. Due to independent dc sources, solar, fuel cells and wind energy system gaining high interest among both in research and industry communities rather than their heavy reliance on changing environmental circumstances [5].

The goal of the current study is to expand inverter reliability by including the Fault Tolerant (FT) characteristic into MLI. Recently, a number of novel MLI topologies with FT capabilities have been presented. Multiple switching conditions that can attain the similar level in the output are necessary for fault-tolerant MLIs. In the literature, these many switching states are referred to as "redundant states" [6]. Because there are fewer "Redundant States" available, the RD-MLI would undoubtedly have a low chance of becoming a fully fault-tolerant architecture. Then, by adding any number of power devices in series or parallel to the current RD-MLI topologies, the redundant path can be achieved. Therefore, a careful adjustment to the current RD-MLI topologies is necessary to make it completely fault-tolerant. Therefore, numerous novel FT MLI have surfaced and been examined in the literature [7].

The issue of dc link voltage balancing with FT RD-MLIs has been examined by certain researchers, who have proposed suitable solutions for defective, post-fault, and normal conditions [7–9]. In Choi et al. [10], a novel approach to OCF identification and justification of 3-level T-type MLI is put forth, with fault finding via measuring capacitor voltage and phase current, which might result in inaccurate finding. Only a partial FT feature is present in the topology of Choi et al. [10], making it not complete FT. Chen et al. [11]'s topology ensures that levels won't drop if a problem occurs and eliminates the need for additional circuitry for voltage regulation of capacitors balancing. However, it needs a number of capacitors and switches than what are required in the proposed SPFT topology. Additionally, the voltage across the capacitor will become unbalanced on the switch if the architecture of Chen et al. [11] is implemented as an FT on any switch that is near a DC link. On the other hand, each switch's faults in the suggested SPFT architecture are acceptable and do not affect voltage balancing.

The literature review suggests that MLI architecture should be completely fault resistant, meaning that they should be able to withstand errors on all of its switches, based on the discussion above. Furthermore, it needs to naturally balance the voltage across DC-Link capacitors in both normal and malfunctioning situations. In this paper focuses on OCF and can perform FT operation assuming the corresponding fault.

This paper is organized as follows: The working principle of the suggested topologies and open circuit fault-tolerant (OCFT) approaches are the main topics of Sect. 2. In Sect. 3, the MATLAB/Simulink outcomes and control methods are explained. Section 4 suggests the experimental confirmation of the suggested SPFT structure. Section 5 presents a comparison of the suggested topologies and their viability. Section 6 concludes with a summary.

2 Working Principle of the Proposed MLI

In this paper, two suggested SPFT multilevel inverter approaches have projected. The extension of the topologies has been presented in the examples.

2.1 Example-1

There is a single-phase packed U-cell MLI topology circuit diagram in Fig. 1 [12]. It uses a 5-level sensor-less U-cell inverter with a self-balancing capacitor for voltage. The single-phase packed U-cell (5-level) MLI is collected of six switches, a constant DC source voltage, and a SVB capacitor. The complimentary scheme controls various switching configurations such as (S_1, S_4), (S_2, S_5) *and* (S_3, S_6). The eight-packed U-cell with five switching state possibilities is listed in Table 1 and is known as $\delta_i(i = 0, 1, 2, 3, 4, 5, 6, 7)$.

Table 2 displays the availability of operating status. As discussed in Vahedi et al. the proposed packed U-cell structure generates a 5-level output voltage wave-shape in normal state. The conducting tracks and current track of the packed U-cell MLI for the 5-level are exposed in Fig. 2.

If any of the switches FS_2, FS_3, FS_5 *and* FS_6 experience an OCF, the architecture can still function with a 3-level output voltage to achieve the packed U-cell MLI. Nevertheless, available states will prevent the synthesis of an alternate MLI output voltage level if any one of the switches $(FS_1\ ,\ FS_4\ )$ experiences an OCF.

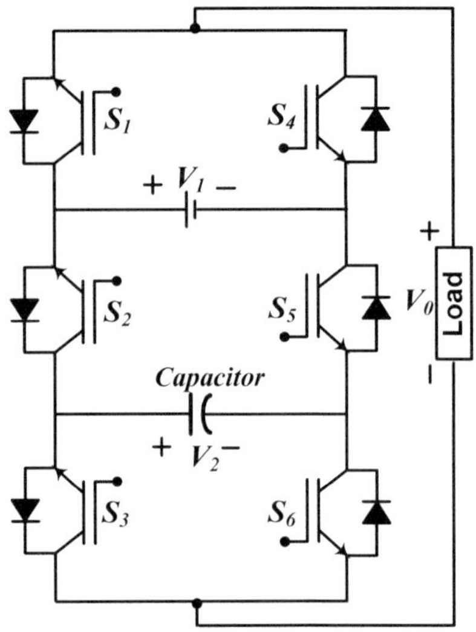

Fig. 1 A typical circuit diagram for a five-level, packed U-cell MLI

Table 1 Possible switching state of the studied packed U-cell (5-level) MLI

State	Switching number	Output voltage magnitude	Output level	Effect on capacitor
δ_0	S_1, S_5, S_6	$+V_1$	$+2E$	Bypassed
δ_1	S_1, S_3, S_5	$+V_1 - V_2$	$+E$	Charged
δ_2	S_1, S_2, S_6	$+V_2$	$+E$	Discharged
δ_3	S_1, S_2, S_3	0	0	Bypassed
δ_4	S_4, S_5, S_6	0	0	Bypassed
δ_5	S_3, S_4, S_5	$-V_2$	$-E$	Charged
δ_6	S_2, S_4, S_6	$V_2 - V_1$	$-E$	Discharged
δ_7	S_2, S_3, S_4	$-V_1$	$-2E$	Bypassed

Table 2 The operational state of each individual switch OCF in a packed U-cell MLI

Failed state	Available states	Active level
FS_1	$\delta_4, \delta_5, \delta_6, \delta_7$	Shut-down
FS_2	$\delta_0, \delta_1, \delta_4, \delta_5$	3-level
FS_3	$\delta_0, \delta_2, \delta_4, \delta_6$	3-level
FS_4	$\delta_0, \delta_1, \delta_2, \delta_3$	Shut-down
FS_5	$\delta_2, \delta_3, \delta_6, \delta_7$	3-level
FS_6	$\delta_1, \delta_3, \delta_5, \delta_7$	3-level

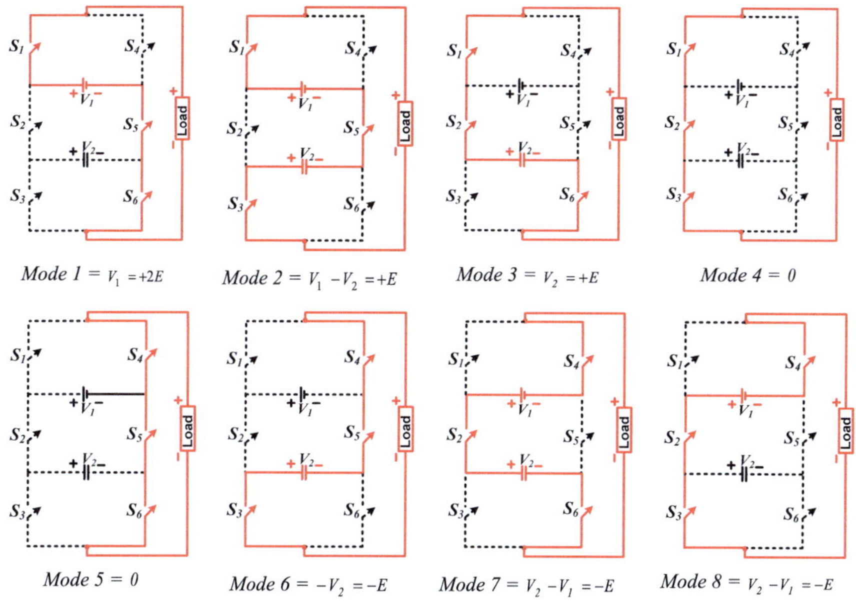

Fig. 2 The packed U-cell MLI's current and conducting paths for the 5-level

2.1.1 Operating Principle of Modified 5-Level MLI

The FT feature blocking the output for switch faults is necessary for the packed U-cell. In order to incorporate intrinsic FT capabilities, the authors have modified the current 5-level packed U-cell MLI. The modified single-phase packed U-cell 5-level MLI with FT functionality, as depicted in Fig. 3, is the result of adding two switches S_7 *and* S_8 to the standard packed U-cell MLI.

The MLI topology with the FT feature and the energy balancing technique between the source and load are the primary topics of this study. Furthermore, the suggested topology's redundancy has grown even further. Table 3 shows the possible switching state for each operational level of the modified packed U-cell (5-levels) MLI in order to take into account the output voltage magnitude.

This is shown in Table 3 in the event that a switch FS_2, FS_3, FS_5, FS_6 fails. The power conversion process continues even though the output is switched from a 5-level to a 3-level and the voltage is lowered. Furthermore, in the event that one of the two switches FS_1 *and* FS_4 experiences OCF, the pre-fault voltage and load current are continued at their pre-fault levels and magnitudes, exactly as they were following the failure.

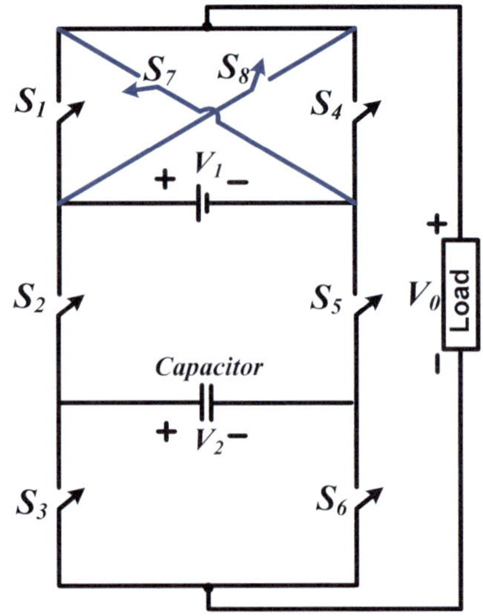

Fig. 3 The suggested SPFT 5-level packed U-cell MLI

Table 3 Operating states of the SPFT modified packed U-cell (5-levels) MLI

State	δ_0	δ_1	δ_2	δ_3	δ_4
ON switches	S_7, S_5, S_6 S_1, S_5, S_6	S_7, S_3, S_5 S_7, S_2, S_6 S_1, S_3, S_5 S_1, S_2, S_6	S_7, S_2, S_3 S_4, S_5, S_6 S_1, S_2, S_3 S_8, S_5, S_6	S_3, S_4, S_5 S_2, S_4, S_6 S_2, S_7, S_6 S_3, S_8, S_5 S_6, S_2, S_8	S_2, S_3, S_4 S_2, S_3, S_8
Output voltage magnitude	$+2E$	$+E$	0	$-E$	$-2E$

2.2 *Example-2*

Figure 4 depicts the Gautam et al. description of the single phase (7-level) symmetrical and asymmetrical multilevel inverter (SA-MLI) construction. Included are two DC voltage sources $V_{S1} and\ V_{S2}$ and capacitors $C_1\ and\ C_2$, which work together to create a voltage divider circuit. Two voltage sources, six unidirectional switches $(S_1, S_2, S_3, S_4, S_5, S_6)$, and a bidirectional switch S_7 make up the SA-MLI (7-level). By connecting the 4-diode $(D_1, D_2, D_3\ and\ D_4)$ to the center of the split DC source, a bidirectional switch is produced.

The given Fig. 4, has consistent switching configurations and states of operation for both SA-MLI (7-level) topologies, and it models the output voltage over a specified load. The output voltage magnitude of both SA-MLI (7-level) devices depends on the switching states and DC voltage. Table 4 lists the several switching configurations $\delta_i (i = 0, 1, 2, 3, 4, 5, 6)$ for the conventional SA-MLI (7-level). Combining

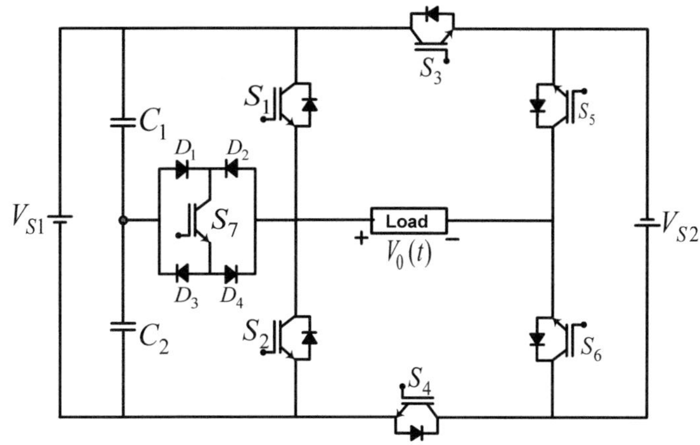

Fig. 4 A 7-level single phase SA-MLI circuit diagram MLI topology

Table 4 7-level SA-MLI circumstances for operation ($V_{S1} = 48\ V,\ V_{S2} = 24\ V$)

State	Switching number	Output voltage magnitude	Output level	Effect on capacitors
δ_0	S_1, S_4, S_5	$+VC_1 + VC_2 + V_{S2}$	$+3E$	Bypassed
δ_1	S_6, S_4, S_1	$+VC_1 + VC_2$	$+2E$	Bypassed
	S_5, S_4, S_7	$+VC_2 + V_{S2}$	$+2E$	C_1 = Charging and C_2 = Discharging
δ_2	S_4, S_7, S_6	$+VC_2$	$+E$	C_1 = Charging and C_2 = Discharging
	S_5, S_4, S_2	$+V_{S2}$	$+E$	Bypassed
δ_3	S_2, S_4, S_6	0	0	Bypassed
	S_1, S_3, S_5	0	0	Bypassed
δ_4	S_1, S_3, S_6	$-V_{S2}$	$-E$	Bypassed
	S_7, S_3, S_5	$-VC_1$	$-E$	C_1 = Discharging and C_2 = Charging
δ_5	S_2, S_3, S_5	$-VC_1 - VC_2$	$-2E$	Bypassed
	S_7, S_3, S_6	$-VC_1 - V_{S2}$	$-2E$	C_1 = Discharging and C_2 = Charging
δ_6	S_2, S_3, S_6	$-VC_1 - VC_2 - V_{S2}$	$-3E$	Bypassed

the sources yielded the magnitude column of the voltage received by the switching combination, as indicated in the table.

The traditional SA-MLI topology produces a seven-level output voltage waveform while operating in normal mode. The current path and directing paths of the SA-MLI topology for the 7-level are displayed in Fig. 5 (Table 5).

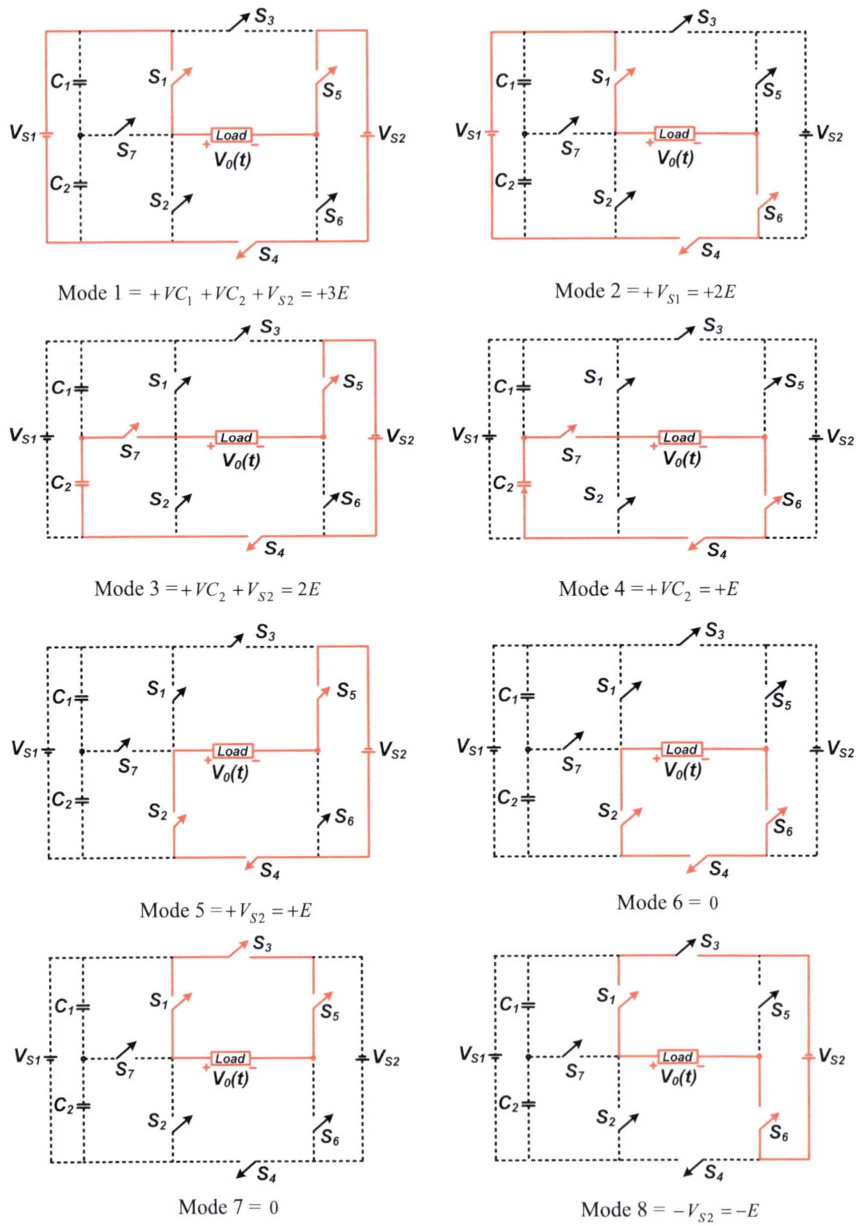

Fig. 5 Current path and conducting paths of MLI for 7-level

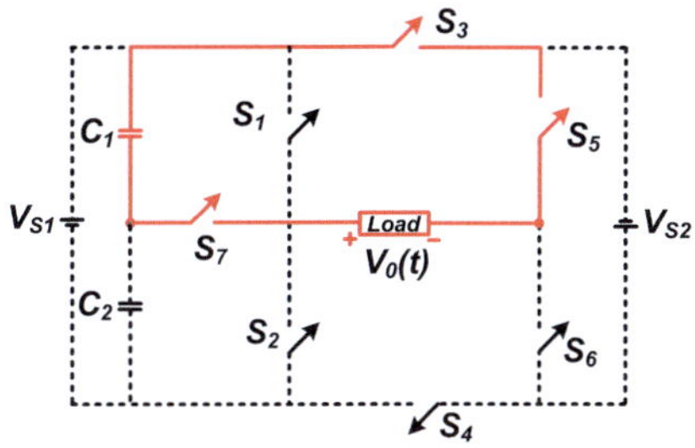

Mode 9 = $-VC_1 = -E$

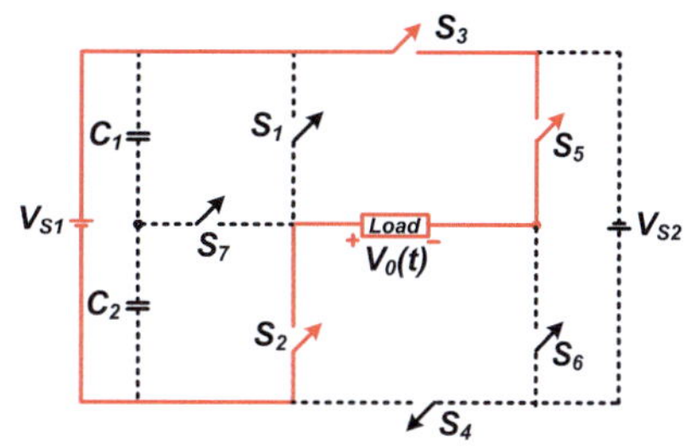

Mode 10 = $-VS_1 = -2E$

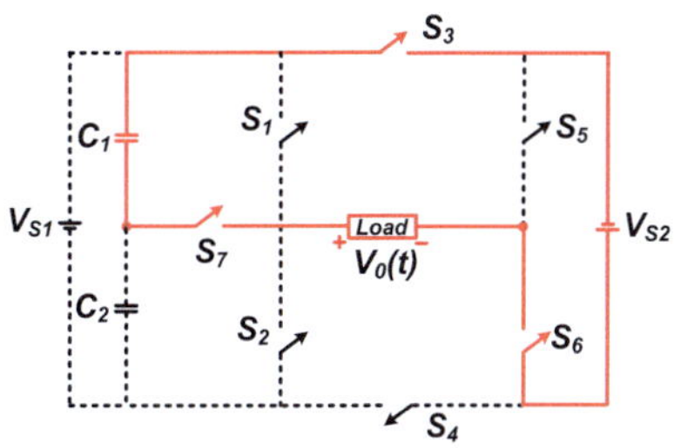

Mode 11 = $-VC_1 - V_{S2} = -2E$

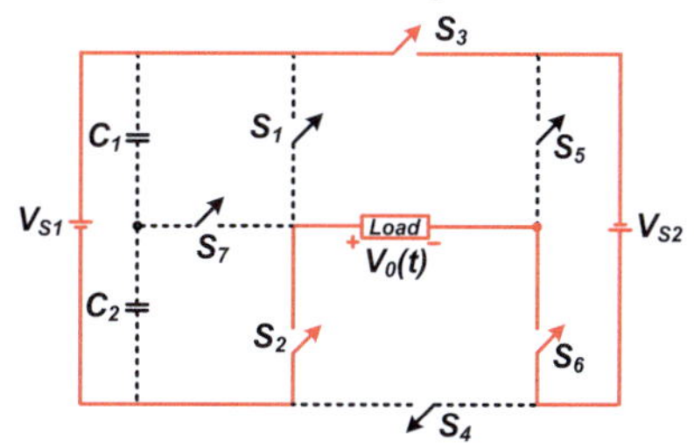

Mode 12 = $-VC_1 - VC_2 - V_{S2} = -3E$

Fig. 5 (continued)

Table 5 Operating position in circumstance of SA-MLI OCF in 7-level

Failed state	Available status	Operation level
S_1	$\delta_1, \delta_2, \delta_3, \delta_4, \delta_5, \delta_6$	5-level
S_2	$\delta_0, \delta_1, \delta_2, \delta_3, \delta_4, \delta_5$	5-level
S_3	$\delta_0, \delta_1, \delta_2, \delta_3$	Shutdown
S_4	$\delta_3, \delta_4, \delta_5, \delta_6$	Shutdown
S_5	$\delta_1, \delta_2, \delta_3, \delta_4, \delta_5$	5-level
S_6	$\delta_0, \delta_1, \delta_2, \delta_3, \delta_4, \delta_5$	5-level
S_7	$\delta_0, \delta_1, \delta_2, \delta_3, \delta_4, \delta_5, \delta_6$	7-level

The architecture can continue to function with a 5-level output voltage in order to achieve the traditional SA-MLI (7-level) in the event that any of the switches (S_1, S_2, S_5, S_6) encounter an OCF. Nevertheless, the operational states will not permit the synthesis of an alternate MLI output voltage level if any one of the switches (S_3 , S_4) experiences an OCF.

2.2.1 Operating Principle of Modified 7-Level MLI

Some changes have been made to the SA-MLI (7-level) to address the issue of open circuit fault (OCF) and make the entire fault bearable. The modified 7-level MLI

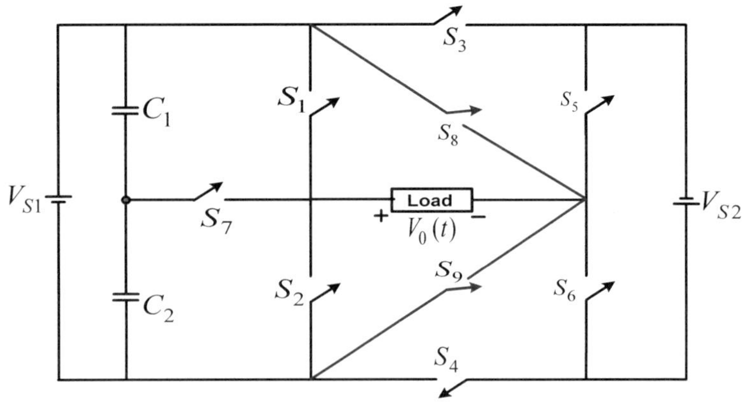

Fig. 6 The proposed modified SPFT 7-level multilevel inverter

architecture seen in Fig. 6 is a generalized version of the normal SA-MLI (7-level) topology, which consists of 9 switches ($S_1, S_2, S_3, S_4, S_5, S_6$ *and* S_7) plus two extra unidirectional switches (S_8 , S_9). Using series-connected DC-link capacitors, the redesigned SA-MLI (7-level) splits the DC source voltage to produce output. Two capacitors connected in series divide its dual DC voltage source. Every capacitor has the same voltage $^{V_{S1}}/_2$.

The MLI topology with the FT feature and the source-load energy balancing system are the primary topics of this study. Furthermore, the suggested topology now has even more redundancy. The output voltage magnitude of the modified SA-MLI (7-level) MLI is taken into consideration by presenting the available switching state for each operational level in Table 6.

In the event that one of the switches (S_3, S_4) fails, Table 6 has shown that. Although the voltage is lower and the output is changed from 7 to 5 levels, the power conversion process is still carried out. Table 7 provides a synopsis of the two suggested examples that were discussed.

3 Simulation Outcomes

This section describes the suggested SPFT topologies' control strategies and simulation outcomes. The suggested FT topology is simulated in the MATLAB/Simulink environment in normal, faulty, and post-fault circumstances. Multicarrier sinusoidal PWM (SPWM) is the method utilized in the suggested SPFT topologies. To create sine modulated PWM pulses for turning on the power switching devices, the multicarrier method that employs the phase opposition disposition PWM (POD-PWM) technique with level shifting (LS-PWM) is used in this study [13]. The proposed scheme is a universal control scheme, in the sense that it can be modulated by any

Table 6 The SPFT adjusted SAMLI (7-level) MLI's operating states

State	δ_0	δ_1	δ_2	δ_3	δ_4	δ_5	δ_6
ON switches	S_1, S_4, S_5	$S_6, S_4, S_1S_5, S_4, S_7$ S_9, S_1	$S_4, S_7, S_6S_5, S_4, S_2$ S_9, S_7	$S_2, S_4, S_6S_1, S_3, S_5S_8, S_1$ S_9, S_2	S_1, S_3, S_6 S_7, S_3, S_5 S_8, S_7	S_2, S_3, S_5 S_7, S_3, S_6 S_2, S_8	S_2, S_3, S_6
Output voltage magnitude	$+3E$	$+2E$	$+E$	0	$-E$	$-2E$	$-3E$

Table 7 Complete summary of the two proposed Fault tolerant SA-MLI

Parameter	Example 1		Example 2	
Levels in normal operation	5-level		7-level	
Faulty switches which cause a complete shutdown	S_1	S_4	S_3	S_4
Lost levels	$+2E, +E$	$-E, -2E$	$-E, -2E, -3E$	$+3E, +2E, +E$
Additional switches	S_7	S_8	S_8	S_9
Available levels	$+2E, +E, 0$	$0, -E, -2E$	$0, -E, -2E$	$+2E, +E, 0$
Levels in the proposed output voltage	5-level		5-level	

output voltage level topology. The schematic diagram of the modulation control strategy is shown in Fig. 7 [14].

In order to obtain the necessary gate pulses, the modulating sinusoidal reference signal ($f_m = 50Hz$) is compared to the 04 triangular carrier signals with carrier or switching frequencies of [14]. The output voltage, load current, and capacitor voltage waveforms are simulated for the suggested SPFT topology, as seen in the figures below. The load under consideration is of the resistive-inductive (RL) variety. For both simulation and experimentation, a modulation index (m_a) of 0.85 and a

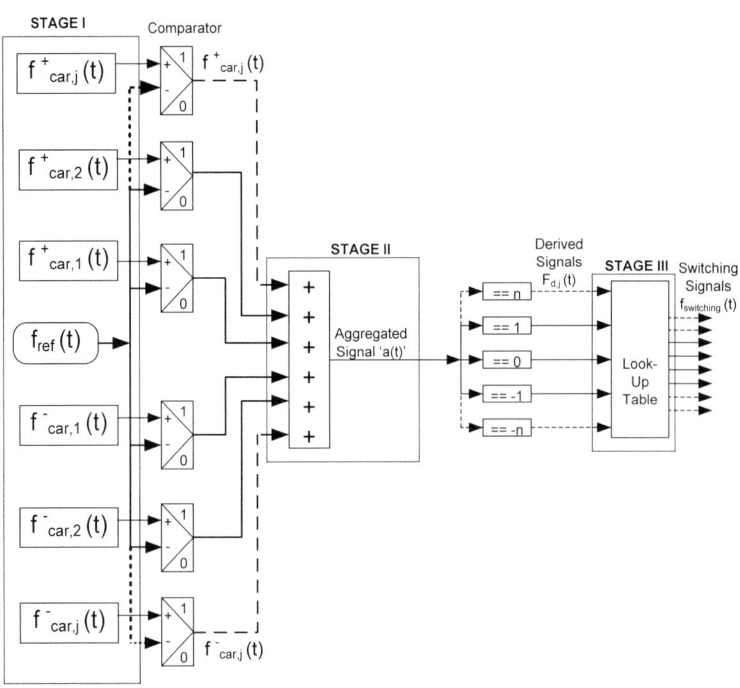

Fig. 7 The schematic diagram of modulation control strategy

modulating wave frequency of (f_m) 50 Hz have been employed. The performance of the suggested topologies is examined by simulation using MATLAB/Simulink and experimental demonstration using an RL load with R = 20 Ω and L = 20 mH, while $C_1 = C_2 = 2500\,\mu F$. Table 8 displays the simulation and experimental parameters for the two suggested SPFT topologies.

Figures 8 and 9 display the simulation outcomes of an OCF created at t = 0.95 s to confirm the FT ability of the packed U-cell (FT 5-level) MLI. Assuming an OCF is generated on the switch at t = 0.95 s, Fig. 8a shows the pre-fault situation for t < 0.95 s and the post-fault condition for t > 0.95 s. As anticipated and illustrated in Figs. 8a and 9a, the 5-level output of the inverter is likewise normal with a peak output of 96V for an applied input voltage of = 96V under normal or pre-fault conditions. Following the creation of an OCF in the switch at t = 0.95 s, the 5-levels in output voltage magnitude and levels prior to pre-fault voltage and load current are maintained to be equal to those following post-fault in order to confirm the FT capability of the suggested SPFT topology.

Figures 10 and 11 display the simulation outcomes of an OCF that was generated at t = 0.94 s in order to confirm the FT ability of the SA-MLI (FT 7-level). The magnitudes for the DC source voltage $V_{S1} = 48\,V \& V_{S2} = 24\,V$ are taken with reference to Fig. 10a. Due to their SVB design, capacitors are thought to be 24V apiece. The estimated voltage rating of each capacitor C_1 *and* C_2 in a potential divider setup is 24V for a rated battery input voltage of 48V.

Table 8 System limitations of proposed MLI topologies

Rated DC voltage (SPFT 5-level)	$V_S = 96\,V$
Rated DC voltage (SPFT 7-level)	$V_{S1} = 48\,V$ and $V_{S2} = 24\,V$
Capacitors	2500 μF
Modulation index m_a	0.85
Fundamental frequency f_m	50 Hz
Switching frequency f_s	3150 kHz
Output load resistance, inductance value	R = 20 Ω, L = 20 mH

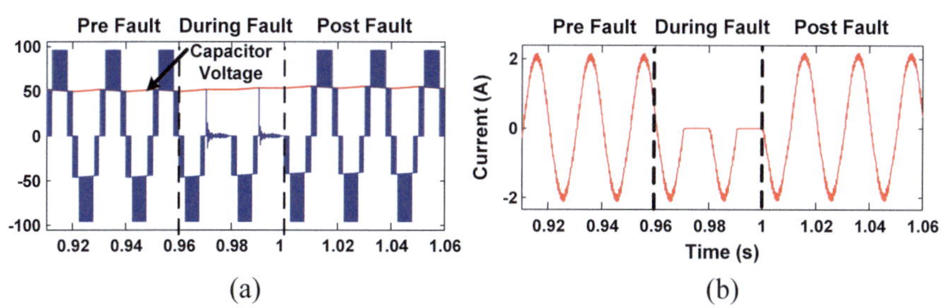

Fig. 8 Simulation outcomes for the proposed SPFT 5-level topology under S_1 OCF under normal, faulty and post-fault state **a** output voltage **b** load current result

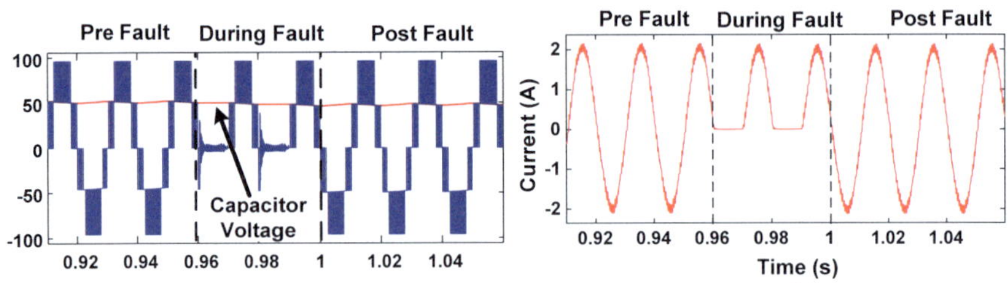

(a) (b)

Fig. 9 Simulation outcomes for the proposed SPFT 5-level topology under S_4 OCF under normal, faulty and post-fault state **a** output voltage **b** load current result

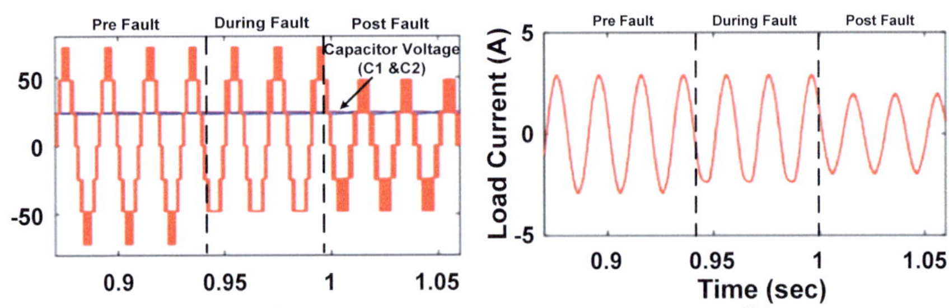

(a) (b)

Fig. 10 Simulation outcomes for the proposed SPFT 7-level MLI under S_3 OCF under normal, faulty and post-fault state **a** Output voltage **b** Load current result

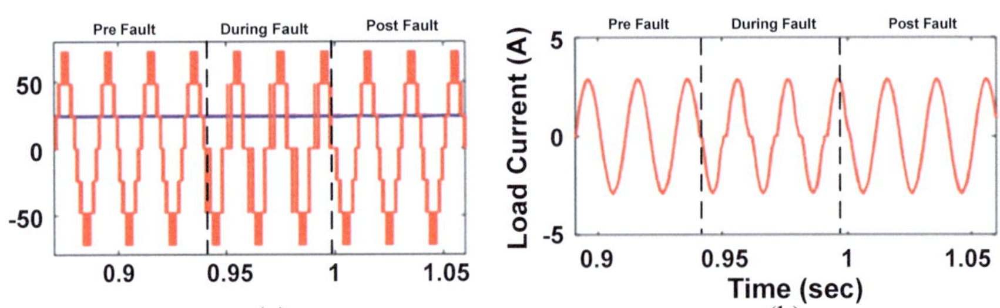

(a) (b)

Fig. 11 Simulation outcomes for the proposed SPFT 7-level MLI under S_7 OCF under normal, faulty and post-fault state **a** output voltage **b** load current result

As seen in Fig. 10a, the output voltage waveform with a 7-level should have a peak-to-peak value of 72 V. Figures 10a and 11a also show the output voltage, load current, and voltage waveforms across the capacitor. The output voltage and current waveforms that were simulated are displayed in Figs. 10 and 11. The suggested SPFT topology's FT capability is then confirmed by creating an OCF in the switch S_3 at t = 0.95 s. As a result, the output voltage's seven levels decrease to five levels occurs while the peak magnitude also decreases. In Fig. 10b, a comparable increase in post-fault load current is also seen. Furthermore, if OCF happens in a switch S_7 , the initial magnitude and levels of the pre-fault voltage and load current are preserved to be identical to those after the post-fault level, as illustrated in Fig. 11.

4 Experimental Validation

An experimental prototype has been created in the lab to confirm that the suggested SA-MLI (FT 7-level) and SPFT packed U-cell (FT 5-level) topologies are represented. Using dSPACE DS-1104 as a real-time controller, control signals are generated for discrete power circuit switching modules (equipped with the proper gate drives M57962L and IGBTs GW30NC120HD). The host computer's MATLAB/Simulink model is interfaced with it. A prototype laboratory setting is depicted in Fig. 12. The scopecorder YOKOGAWA DL750E is used to record and measure all of the testing data. We have measured the output voltage, current, and THD using the Fluke 43B power quality analyzer.

The SPFT suggested topologies' performance is examined by simulation using MATLAB/Simulink and experimental demonstration using an RL load with R = 20Ω and L = 20 mH, while $C_1 = C_2 = 2500\,\mu F$. outcomes from experiments implementing the suggested topologies (5-level and 7-level) in terms of voltage, load current, and THD across the RL load are shown in Figs. 13 and 14. Furthermore, Figs. 13a and 14a show the output voltage, load current, and voltage waveforms across the capacitor.

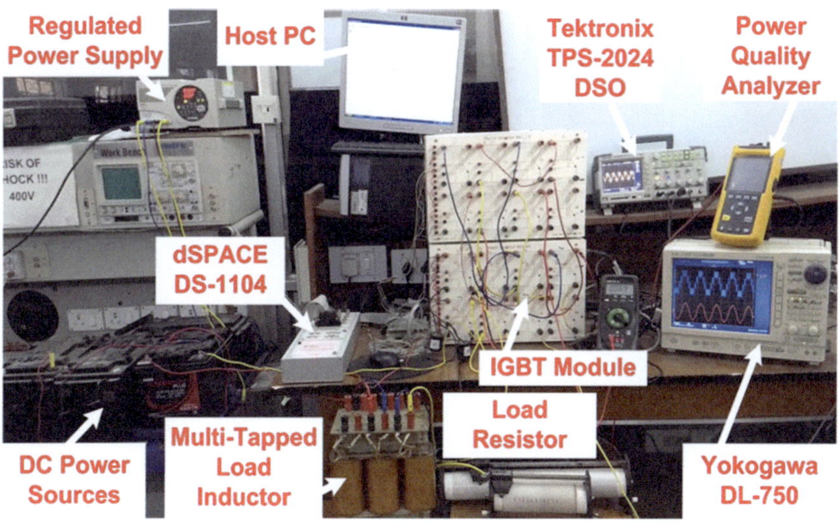

Fig. 12 Experimental setup [15]

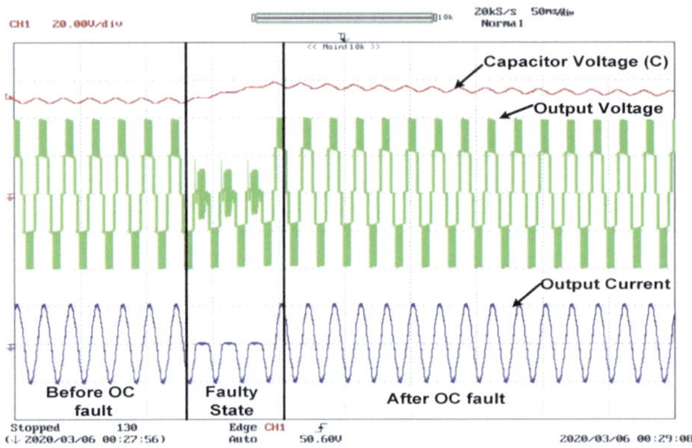

(a)

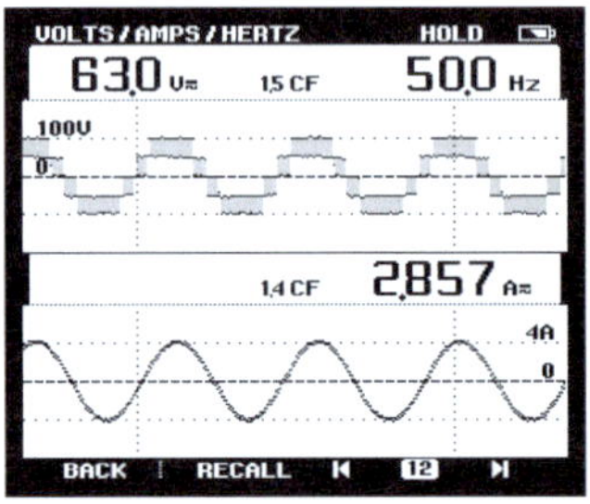

(b)

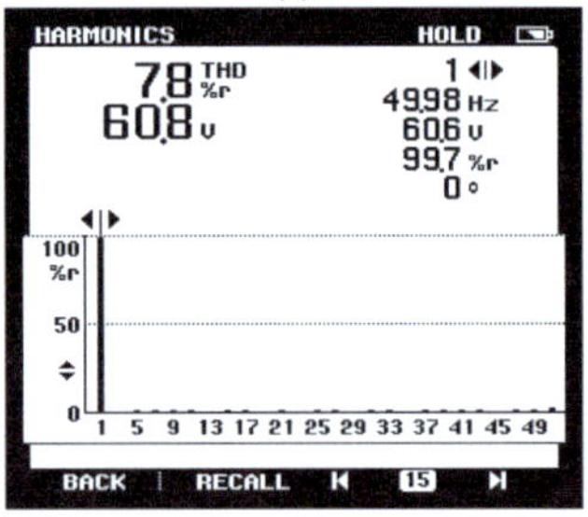

(c)

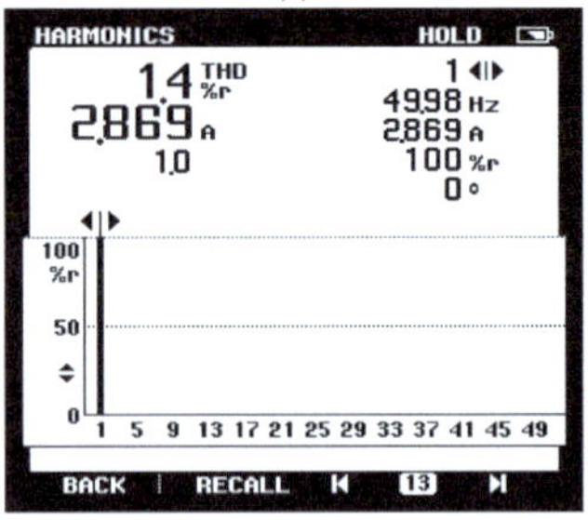

(d)

Fig. 13 Experimental outcomes of the proposed SPFT 5-level MLI under OCF states **a** Output voltage, load current and capacitor voltage ripples waveforms under normal, faulty and post-fault conditions **b** load voltage and current profile under normal state **c** voltage harmonic spectrum under normal state **d** current harmonic spectrum result under normal state

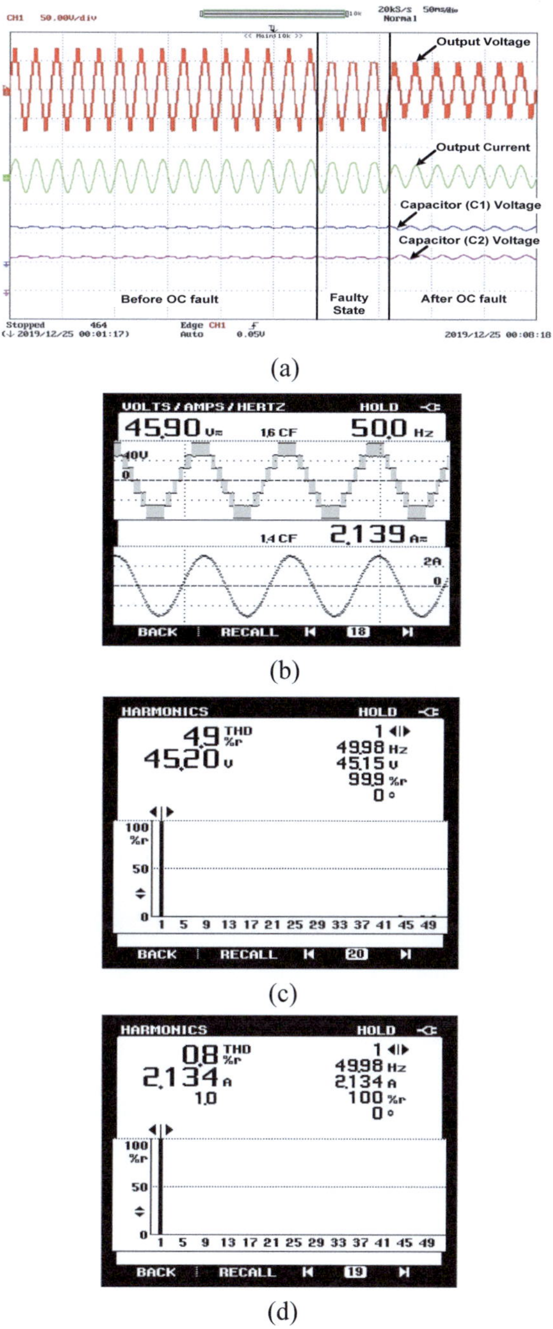

Fig. 14 Experimental outcomes of the proposed SPFT 7-level MLI under OCF states **a** output voltage, load current and capacitor voltage ripples waveforms under normal, faulty and post-fault states **b** load voltage and current profile under normal state **c** Voltage harmonic spectrum under normal state **d** current harmonic spectrum result under normal state

Likewise, an OCF has been developed for a switch in the proposed FT 7-level architecture; Fig. 14 illustrates the power conversion that continues with at least a 5-level waveform. Before OCF conditions, Fig. 14 displays the testing findings for the SPFT planned 7-level output voltage of 72V peak to peak and 45.9V RMS prototype MLI output at a frequency of 50Hz. Figure 14c and d show the suggested 7-level, which satisfies the IEEE-519 harmonics standard with a voltage THD of 4.9% and a current THD of 0.8%. A module has also been tested with RL load, and it performed well. The suggested topologies have been shown to provide the FT feature for the modified SPFT 5-level and 7-level MLIs topologies in a satisfactory manner.

5 Modified MLI Topologies and Existing MLI Topologies Are Compared

This section compares the topologies and the pros and disadvantages of the necessary parameters. Existing RD-MLI topologies have been compared with the suggested modified SPFT 5-level and 7-level topologies in Tables 9 and 10. Dependability, FT ability, and the no. of DC sources, capacitors, drivers, diodes, switches, and bidirectional switches are among the crucial criteria that have been examined.

According to Table 9, all of the suggested topologies display FT characteristics (5-level). Compared to the other topologies analyzed in Table 9, the suggested SPFT topology has higher dependability since it necessitates RD-MLI.

Table 10 indicates (7-level), the comparison of the MLI parameters in terms of power electronic drives to generate a 7-level output voltage.

This is an important advantage of modified MLI topologies because capacitors are an important factor in reducing the bulkiness of MLI. Furthermore, in contrast to other topologies, the suggested SPFT topologies call for the reduced device (RD) of the diode, drivers, and IGBTs.

6 Conclusion

In order to solve the OCF in switches with fewer devices, this research suggests single phase fault tolerant (SPFT) MLI topologies with many redundant levels. Therefore, two examples have been taken for a view of imparting fault tolerant (FT) operation to them. Both the examples (5-level and 7-level) in their initial form do not sustenance the fault-tolerant function. Presented in this paper is an optimal addition of one switch in 5-level and two switches in 7-level have added to make the complete fault-tolerant. DC-link capacitors can automatically maintain SVB under normal, defective, and post-fault process with the help of the suggested SPFT MLI topologies. Under all suggested FT operations' operating settings, the control scheme is run utilizing the LS-PWM approach. The operation of the topology principle has been demonstrated

Table 9 Comparison of 5-level MLI topologies

Topology	DC sources	Capacitors	Diodes	Uni-directional switches	Bi-directionalswitches	FT/reliability (YES/NO)
CHB	2	0	0	8	0	Y
Chen et al. [11]	1	8	0	22	0	Y
Nicolas-Apruzzese et al. [16]	4	0	0	20	0	Y
Sivakumar et al. [17]	2	0	2	7	1	Y
Gautam et al. [18]	1	2	2	7	1	Y
Dewangan et al. [19]	2	0	0	6	1	Yes
Proposed topology	1	1	0	8	0	Yes

Table 10 Comparison of existing 7-level MLI

Topology	DC sources	Capacitors	Diodes	Uni-directional switches	Bi-directional switches	FT/ reliability (YES/NO)
CHB	3	0	0	12	0	Y
Hsieh et al. [20]	1	3	4	8	0	N
Sun et al. [21]	1	3	0	12	2	N
Choi et al. [22]	1	3	2	7	0	N
Lee et al. [23]	1	2	0	12	2	N
Haji-Esmaeili et al. [24]	3	0	0	12	0	Y
Choupan et al. [25]	3	0	0	8	2	Y
Proposed topology	2	2	0	8	1	Y

by the outcomes. Therefore, the outcomes assert that the recommended SPFT MLI topologies have fault-tolerant capabilities.

References

1. Zhang W, Xu D, Enjeti PN, Li H, Hawke JT, Krishnamoorthy HS (2014) Survey on fault-tolerant techniques for power electronic converters. IEEE Trans Power Electron 29(12):6319–6331
2. Rodriguez J, Lai JS, Peng FZ (2002) Multilevel inverters: a survey of topologies, controls, and applications. IEEE Trans Industr Electron 49(4):724–738
3. Malinowski M, Gopakumar K, Rodriguez J, Perez MA (2009) A survey on cascaded multilevel inverters. IEEE Trans Industr Electron 57(7):2197–2206
4. Kouro S, Malinowski M, Gopakumar K, Pou J, Franquelo LG, Wu B, Rodriguez J, Pérez MA, Leon JI (2010) Recent advances and industrial applications of multilevel converters. IEEE Trans Ind Electron 57(8):2553–80
5. Eroğlu F, Kurtoğlu M, Arslan AO, Vural AM (2019) Harmonic reduction under unbalanced operating conditions of PV-connected cascaded H-bridge multilevel inverters using fault tolerant adaptive phase-shifted pulse width modulation. Int Trans Electr Energy Syst 29(4):e2814
6. Dewangan NK, Gupta S, Gupta KK (2019) Fault-tolerant operation of some reduced-device-count multilevel inverters with improved performance. Int Trans Electr Energy Syst 29(2):e2731
7. Raj N, Kale T, Anand A, George S (2018) Switch fault detection and diagnosis in space vector modulated cascaded H-bridge multilevel inverter. Int J Electron 105(12):1977–1992
8. Jalhotra M, Sahu LK, Gupta S, Gautam SP (2019) Resilient fault tolerant topology of single phase multilevel inverter. IEEE J Emerg Sel Top Power Electron
9. Yazdani A, Sepahvand H, Crow ML, Ferdowsi M (2010) Fault detection and mitigation in multilevel converter STATCOMs. IEEE Trans Industr Electron 58(4):1307–1315

10. Ceballos S, Pou J, Robles E, Zaragoza J, Martin JL (2009) Performance evaluation of fault-tolerant neutral-point-clamped converters. IEEE Trans Industr Electron 57(8):2709–2718
11. Choi UM, Lee KB, Blaabjerg F (2013) Diagnosis and tolerant strategy of an open-switch fault for T-type three-level inverter systems. IEEE Trans Ind Appl 50(1):495–508
12. Chen A, Hu L, Chen L, Deng Y, He X (2005) A multilevel converter topology with fault-tolerant ability. IEEE Trans Power Electron 20(2):405–415
13. Abarzadeh M, Vahedi H, Al-Haddad K (2019) Fast sensor-less voltage balancing and capacitor size reduction in PUC5 converter using novel modulation method. IEEE Trans Industr Inf 15(8):4394–4406
14. Santosh Kumar M, Borghate VB, Karasani RR, Sabyasachi S, Suryawanshi HM (2018) A fault-tolerant modular multilevel inverter topology. Int J Circuit Theory Appl 46(5):1028–1043
15. Gupta KK, Jain S (2013) A novel multilevel inverter based on switched DC sources. IEEE Trans Industr Electron 61(7):3269–3278
16. Nicolas-Apruzzese J, Busquets-Monge S, Bordonau J, Alepuz S, Calle-Prado A (2012) Analysis of the fault-tolerance capacity of the multilevel active-clamped converter. IEEE Trans Industr Electron 60(11):4773–4783
17. Sivakumar K (2015) A fault-tolerant single-phase five-level inverter for grid-independent PV systems. IEEE Trans Industr Electron 62(12):7569–7577
18. Gautam SP, Gupta S, Kumar L (2017) Reliability improvement of transistor clamped H-bridge-based cascaded multilevel inverter. IET Power Electronics 10(7):770–781
19. Dewangan NK, Gupta S, Gupta KK (2018) Approach to synthesis of fault tolerant reduced device count multilevel inverters (FT RDC MLIs). IET Power Electronics 12(3):476–482
20. Hsieh C, Liang T, Chen S et al (2016) Design and implementation of a novel multilevel DC-AC inverter. IEEE Trans Ind App 52(3):2436–2443
21. Sun X, Wang B, Zhou Y, Wang W, du H, Lu Z (2016) A single dc source cascaded seven-level inverter integrating switched-capacitor techniques. IEEE Trans Ind Electron 63(11):7184-7194
22. Choi J, Kang F (2015) Seven-level PWM inverter employing series-connected capacitors paralleled to a single DC voltage source. IEEE Trans Ind Electron 62(6):3448–3459
23. Lee SS (2018) A single-phase single-source 7-level inverter with triple voltage boosting gain. IEEE Access 6:30005–30011
24. Haji-Esmaeili MM, Naseri M, Khoun-Jahan H, Abapour M (2017) Fault-tolerant structure for cascaded H-bridge multilevel inverter and reliability evaluation. IET Power Electron 10(1):59–70
25. Choupan R, Golshannavaz S, Nazarpour D, Barmala M (2019) A new structure for multilevel inverters with fault-tolerant capability against open circuit faults. Electr Power Syst Res 168:105–116

Machine Learning Based Fault Detection and Classification in Multilevel Inverter for Industrial Applications

Niraj Kumar Dewangan, N. D. Jeevan, Vivek Gurjar, Kasinath Jena, and Dhananjay Kumar

Abstract Multilevel Inverters (MLIs) are widely used in commercial applications that demand high voltage and power levels. An increase in switch count in the MLI leads to higher likelihood of component failures. Detecting switch faults in power converters is therefore crucial. This study focused specifically on diagnosing open-circuit (OC) switch faults in MLIs. The proposed fault diagnosis (FD) technique is depending on machine learning and relies on output voltage data. From this voltage, three key features were extracted such as Total Harmonic Distortion (THD), Root Mean Square (RMS) value and Mean value. The Machine Learning (ML) algorithms used for diagnosis of OC switch faults were, Decision Trees (DT), Support Vector Machines (SVM), Random Forests (RF), and K-Nearest Neighbors (KNN). The diagnostic framework was developed and tested in the MATLAB/Simulink tool. Among the models, DT achieved maximum accuracy of 99.76% with a training and testing data split of 70% and 30% respectively.

N. K. Dewangan (✉) · N. D. Jeevan
Department of Mechatronics, Manipal Institute of Technology, Manipal Academy of Higher Education, Manipal, India
e-mail: niraj.dewangan@manipal.edu

N. D. Jeevan
e-mail: jeevan.mitmpl2024@learner.manipal.edu

V. Gurjar
Department of ECE, Master of Science, Cleveland State University, Cleveland, USA
e-mail: v.gurjar@alumni.csuohio.edu

K. Jena
Department of Electrical and Electronics Engineering, School of Engineering and IT, ARKA JAIN University, Jharkhand, India
e-mail: kasi.jena@gmail.com

D. Kumar
Department of Electrical Engineering, Government Engineering College, Siwan, Bihar, India
e-mail: dhananjaymanit23@gmail.com

Y. Gopal et al. (eds.), *Multilevel Inverter Technology for Renewable Energy Systems*, Green Energy and Technology, https://doi.org/10.1007/978-3-032-03589-9_9

1 Introduction

In power industry, Multilevel Inverters (MLIs) received increased prominence in power electronics for DC/AC power conversion application due to their better performances in terms of superior quality of power, reduced voltage stress on switches, improved electromagnetic compatibility, and decreased use of bulk passive filters compared to two-level power converters. All these advantages resulted in MLIs being one of the preferred choices in various emerging applications like power systems, active power filters, Electric Vehicle (EV) drives, Industrial Drives (ID), and renewable energy systems [1–5]. With more semiconductor switches in MLI configurations, there is increased complexity and an increased chance of switch faults occurring. Among these switch faults, open circuit (OC) and short circuit (SC) faults are most critical ones. It is indicated that about 21% of faults of MLIs occur due to these switch faults [6], which lead to higher Total Harmonic Distortion (THD) content, reduced efficiency, and lowered reliability.

In light of these issues, fault detection and diagnosis at an early stage is crucial to safeguarding system functionality and avoiding devastating faults. Several methods have been designed to be used for this, which are model based, signal-processing based, and databased methods [7]. Model-based methods assume an exact mathematical model of the system, which becomes very difficult to establish as the system increases in complexity. Signal-processing-based methods, require extra hardware to condition the signals. Data-based methods, provide an adaptable and scalable solution, based on history of data. The data based methods include intelligent methods like Machine Learning (ML) and Deep Learning (DL) and fuzzy logic [7]. Several switch fault detection methods for multilevel inverters have been proposed in the literature using ML, DL, and Artificial Neural Network (ANN) algorithms with various feature extraction schemes. A twenty-one-level MLI using a lightweight shuffle-Sim AM network for fault detection was proposed in [8]. A fault identification method identifying faults in modular multilevel converters as functions of variations in capacitor voltage was proposed in [9]. A hybrid diagnosis method using ANN for open-switch fault detection and Discrete Wavelet Transform (DWT) for feature extraction of output voltages and current was proposed in [10]. Various open-switch fault detection and localization methods were discussed in [10–13]. Techniques for detecting and localizing open-switch faults were investigated in [14] using various ML algorithms like Logistic regression (LR), Decision Trees (DT), AdaBoost, KNN, Voting Classifier (VC), and XGBoost, wherein VC is to be an appropriate classifier for fault diagnosis (FD). In [15], occurrence of twenty different faults in solar-powered cascaded H-bridge MLI was demonstrated. An open fault detection method using Back-Propagation Neural Network (BPNN) in seven-level cascaded H-bridge inverter was presented in [16] using harmonic features of output voltage to train the neural network. An automated fault classification method using multiscale kernel convolutional neural network (MKCNN) based on deep learning was proposed for single or multiple OC, and SC switch faults in [17]. In this method, simulation data were converted into images using Short-Time Fourier Transform (STFT) to extract important features. The FD method using Probabilistic Principal Component Analysis (PPCA) and ML algorithms was studied in [18]. The application of XGBoost algorithm as means of identifying OC switch faults in multilevel modular converters (MMCs) was explored in [19]. An open circuit fault (OCF) detection and identification method for three-phase, Photo-Voltaic (PV) fed neutral point clamped (NPC) inverter on switches using ML classification models was discussed in [20]. Recursive Discrete Fourier Transform (RDFT) and DWT were utilized as methods of feature extraction.

The research focuses on FD of OCFs in MLIs based on ML methods. The originality of this work is no research has been conducted on switch FD for the MLI presented in [21]. The ML algorithms utilized are KNN, Support Vector Machine (SVM), DT, and Random Forest (RF). The algorithms exhibit high accuracy and precision in identifying power switch OC faults. The fault detecting process only depends on measuring the output voltage. Feature extraction is also necessary before these diagnostic techniques are applied.

2 Five-Level Multilevel Inverter: Fault Analysis

2.1 Topology of MLI

The MLI topology, illustrated in Fig. 1, consists of six IGBT switches with an antiparallel diode and two DC sources. The switches S_1, S_1', S_2, S_2', S_3, and S_3' are connected as shown in the figure. The switches in the two legs operate in a complementary manner, when switch S_1 is ON, S_1' is OFF, and similarly for the remaining switches. Table 1 shows the different switching combinations used to obtain the individual levels for the MLI topology.

2.2 Switching Strategy

The modulation technique employed in this study is Level-Shifted Pulse-Width Modulation (LS-PWM). In this scheme, the modulating wave is continuously compared with triangular carrier waves to obtain appropriate gating pulses for the switches, as illustrated in Fig. 2. Simulations are performed using the MATLAB/Simulink platform. The simulation model consists of a MLI connected to an RL load. The system parameters are provided in Table 2. Figure 3 displays waveform of the output voltage under normal operating conditions.

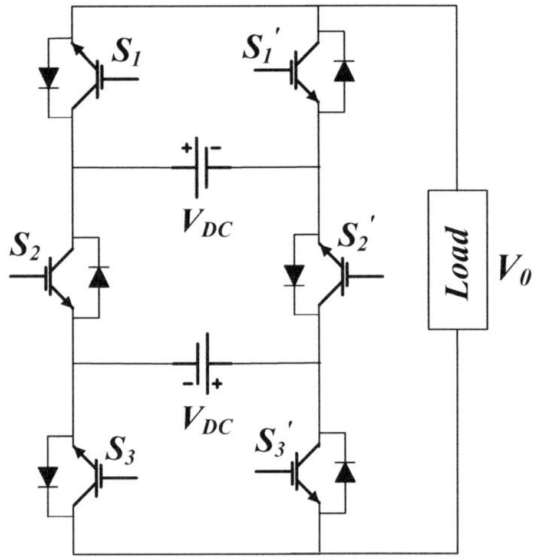

Fig. 1 Five-level MLI topology [21]

Table 1 Switching states for five-level MLI topology [21]

Modes	Switch conditions			Corresponding output voltage
	S_1	S_2	S_3	
1	0	1	0	$-2V_{DC}$
2	0	1	1	$-V_{DC}$
3	1	1	0	
4	1	1	1	0
5	0	0	0	
6	1	0	0	$+V_{DC}$
7	0	0	1	
8	1	0	1	$+2V_{DC}$

* '1' stands for ON state of switch and '0' stands for OFF state of the switch *

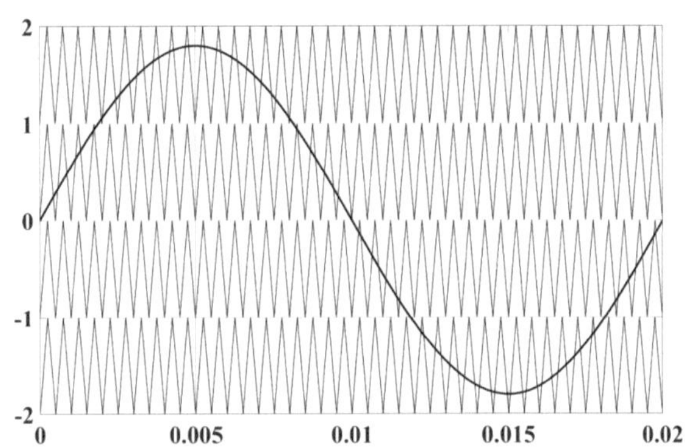

Fig. 2 LS-PWM generation technique

Table 2 Specification of the MLI model

Parameter		Specification
Source voltage (DC)		100 V
Load	Resistance	5 Ω
	Inductance	2mH
Modulating signal frequency		50 Hz
Carrier wave frequency		2.5 kHz

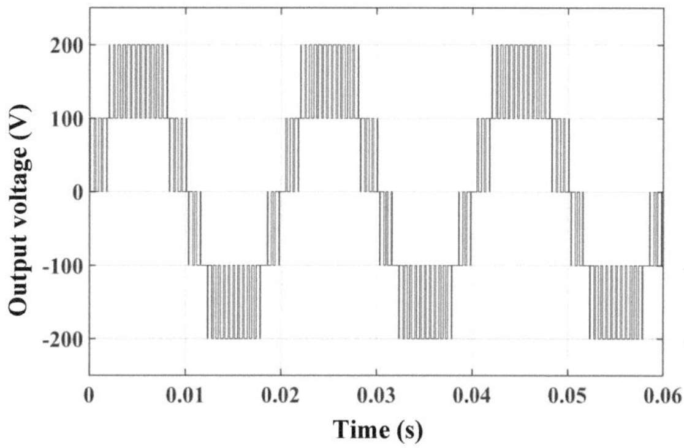

Fig. 3 Waveform of output voltage in normal conditions

2.3 MLI OCF Analysis

Two major kinds of switch faults in MLI modules are OCFs and SC faults. In OCF conditions, the switch will not conduct and cannot pass current to the load. During SCF conditions, excessive current flows, making interruption difficult. OCF identification is the main focus of the paper. In this paper only single OCF in switches are considered. When an OCF occurs on the switches the waveform shape is degraded. Suppose an OC faults on switches, S_1, S_1', S_2, S_2', S_3, and S_3' are created at $t = 0.02$ s, the corresponding voltage waveforms are shown in Fig. 4.

3 Open Circuit Fault Classification Technique

3.1 Flow Chart for Fault Classification

Figure 5 presents a flowchart of the OC fault classification method. The MLI consists of cross connected DC sources and connected with RL load for the fault analysis. The MLI output voltage is considered for the FD. From the MLI output voltage features should be extracted for the proper classification of fault, otherwise misclassification takes place, because of similar waveforms for multiple individual switch faults. This

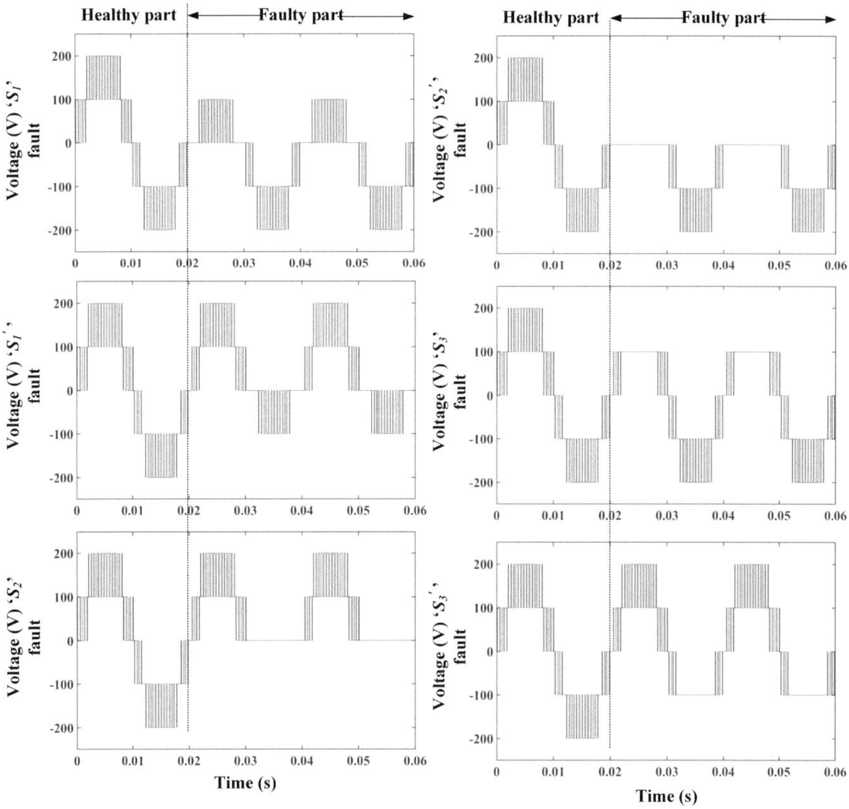

Fig. 4 Output voltage waveforms under faulty conditions, fault initiated at t = 0.02 s (for S_1, S_1', S_2, S_2', S_3, and S_3')

paper, considers THD, Root Mean Square (RMS) and Mean of voltages as features. Then different ML algorithms needs to be applied. Here in this step each ML model is trained and tested for the data extracted. Finally applying the best classifier for accurate classification of OC switch faults.

3.2 Feature Extraction (FE) Technique

FE is an essential step in diagnosing switch faults. The proposed study uses three features of output voltage: THD, RMS and mean of voltage.

The THD can be mathematically expressed as,

$$THD = \frac{\sqrt{V_2^2 + V_3^2 + V_4^2 + \ldots + V_n^2}}{V_1}$$

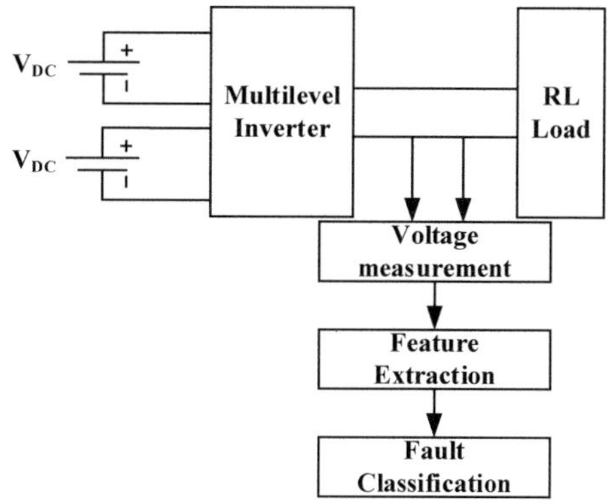

Fig. 5 Flowchart of FD method

where, $V_1, V_2, V_3 \ldots V_n$ are the RMS values of the fundamental frequency and harmonics components of the output voltage waveform, respectively.

The RMS value of voltage can be expressed as,

$$V_{RMS} = \sqrt{\frac{1}{T}\int_0^T v_o(t)^2 dt}$$

where,$v_o(t)$ is the instantaneous value of output voltage as a function of time. "T" is the time period.

The Mean value of voltage can be represented as,

$$V_{mean} = \sqrt{\frac{1}{T}\int_0^T v_o(t) dt}$$

3.3 *Different ML Models Used for the FD*

The proposed research evaluates the effectiveness of four ML algorithms in diagnosing OCFs in inverter switches. The effectiveness of each ML model is analyzed based on specific metrics.

(a) Decision Trees (DT): A Decision Tree is a type of Supervised Learning (SL) method that utilizes a branched structure to solve both Classification and Regression (C&R) problems. It works by repeatedly dividing the total dataset into smaller datasets based on specific feature values, creating branches at each decision point, and ending with leaf nodes that represent the final outcomes. The algorithm chooses the best splits at each step using measures like Information Gain, Gini Impurity, or Mean Squared Error. DT is easy to understand and interpret, and it can handle both types of data: numerical as well as categorical.

(b) Support Vector Machine (SVM): It is an effective SL method used for both C&R tasks. It works by identifying the best possible hyperplane that separates different groups of data with the widest margin. The data points closest to this boundary called support vectors are key in shaping the decision line or surface. SVM uses kernel functions like linear, polynomial, etc to convert the input data into higher-dimensional spaces so that it can handle both linear and non-linear data. While SVM delivers strong performance, especially in complex, high-dimensional scenarios, it can be resource-intensive for large datasets and often requires careful tuning of parameters like the type of kernel and the strength of regularization.

(c) K-Nearest Neighbors (KNN): KNN is a SL based algorithm used for both C&R tasks. It classifies a new data point by looking at the "k" nearest training examples and assigning the most common class among them, usually based on distance typically Euclidean. Choosing the right value of "k" is important for performance, and odd values are often used to avoid ties during classification. In this work, the 'k' value is set to 5. KNN is somehow easy to implement, but it can be affected by irrelevant features and tends to be slow for the predictions on large datasets.

(d) Random Forest (RF): It is an ensemble learning technique that uses multiple DT for the prediction accuracy. It uses bootstrap sampling to generate a subgroup of training data for each tree and introduces randomness by choosing a random subgroup of features at each node. The final prediction is determined by the highest vote in classification tasks or by averaging the values in regression tasks. This method significantly decreases the risk of overfitting with respect to individual trees. While Random Forests deliver high performance and are resilient to noisy data, they tend to be less interpretable and computationally heavier than single decision tree models.

4 Results and Discussion

To validate the presented fault diagnostic approach, simulations were conducted using the MATLAB/Simulink platform. In this process, only the output voltage was monitored and used for data acquisition. The voltage signals were directly exported to the MATLAB workspace to gather a comprehensive dataset suitable for ML classification. The dataset, which includes both normal (healthy) and fault conditions, is summarized in Table 3. From the acquired output voltage data, key features such as THD, RMS, and Mean voltage were extracted. These features serve as the input to the ML models, while the output is a fault label that gives the location of the fault. For instance, if an OCF occurs in switch S_1, the ML classifier outputs label '1'; if the fault is in switch S_2, the label is '2', and so on, as detailed in Table 5. A healthy condition is represented by the fault label '0'.

Table 3 Dataset prepared for MLI OC switch fault classification

Number of labels	7
Total data size	201 × 7
Training data size	140 × 7
Testing data size	61 × 7

Table 4 Performance metrics of different ML classifier

ML classifier	Accuracy (%)	Precision (%)	Recall (%)	F1 score (%)
DT	99.76	99.82	99.75	99.78
SVM	99.52	99.57	99.53	99.55
KNN	99.29	99.35	99.29	99.31
RF	99.52	99.56	99.54	99.54

Table 5 Fault labels, fault detection and fault classification

Switch fault	Detection time in seconds	Classification time in seconds	Fault label
S_1	0.024	0.0288	1
S_2	0.0338	0.0364	2
S_3	0.0328	0.036	3
S_4	0.0228	0.0288	4
S_5	0.0245	0.0255	5
S_6	0.0352	0.0356	6
No fault	–	–	0

The primary target of this study is to examine the effectiveness of ML classifiers in identifying and classifying faults based on the collected data. The ML algorithms evaluated in this work include DT, SVM, RF, and KNN. Each model was evaluated using raining and testing data split of 70% and 30% respectively. The classifiers were evaluated based on key performance metrics: accuracy, precision, recall, and F1 score, as shown in Table 4. The confusion matrices for each algorithm are presented in Fig. 6. Based on the analysis, the DT model achieved the highest performance, with an accuracy of 99.76%, precision of 99.82%, recall of 99.75%, and F1 score of 99.78%.

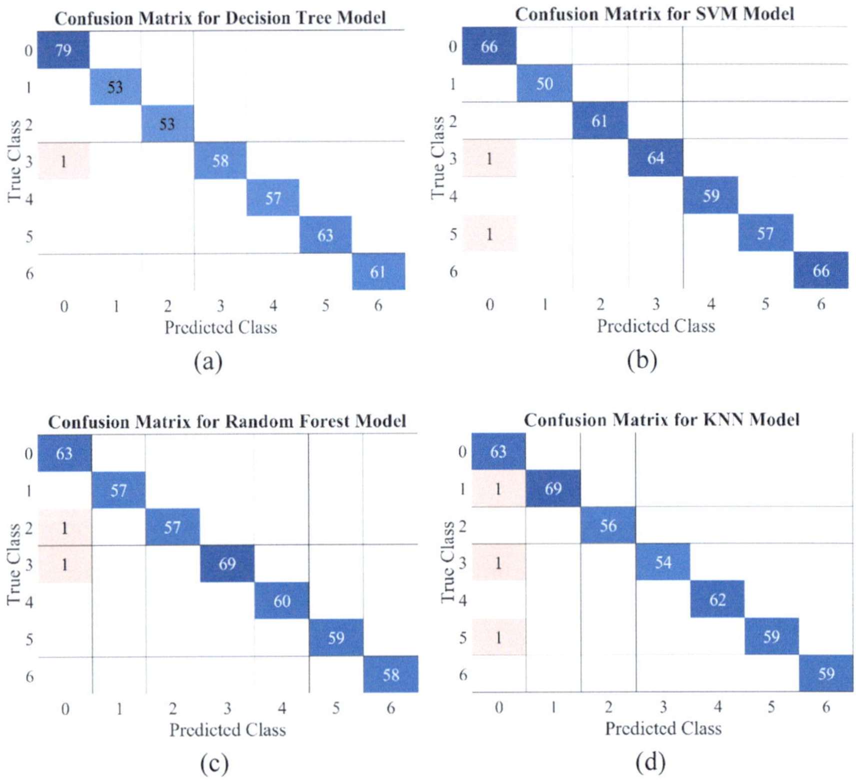

Fig. 6 Confusion matrices of **a** DT, **b** SVM **c** RF **d** KNN

The subsequent step involves assessing the performance of the best-performing classifier, DT, specifically in identifying and classifying OC switch faults. To examine the classification accuracy of the DT model, a simulation was carried out for an OC fault occurring in switch S_3. As illustrated in Fig. 7, the output voltage waveform, fault initiation time, detection time, and classification time are presented for the fault in switch S_3. The DT model successfully detected the OC fault within 0.0328 s and accurately classified it as a fault in S_3 at time $t = 0.036$ s. Similar simulations were performed individually for each switch, with OC faults introduced at 0.02 s. The corresponding detection and classification times for all switches are summarized in Table 5.

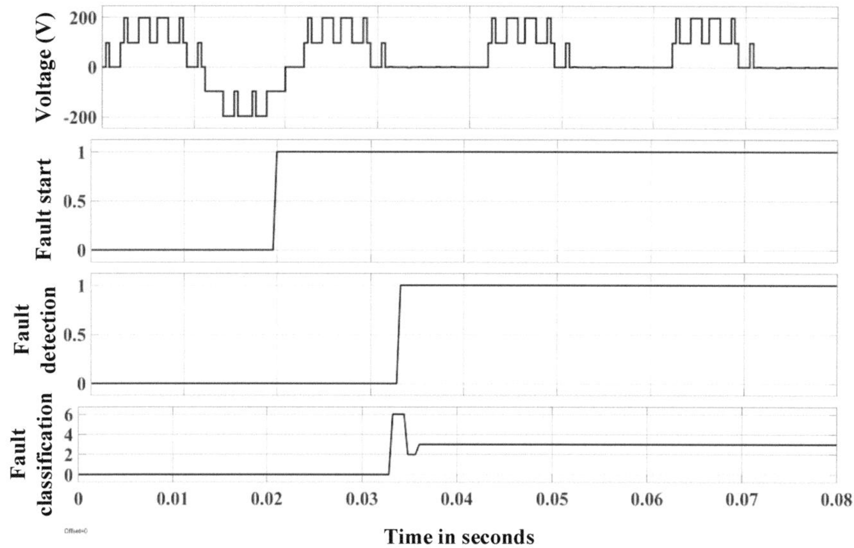

Fig. 7 Simulation results of switch S_3 **a** output voltage waveform under faulty condition, **b** fault initiated at $t = 0.02s$, **c** fault detected at $t = 0.0328s$, **d** fault classified at $t = 0.036s$

5 Conclusion

The proposed study effectively identified single-switch OCFs using a ML based approach. The fault detection strategy relied solely on the output voltage, from which key features such as THD, RMS, and Mean were extracted. These features served as inputs to four ML algorithms: DT, SVM, RF, and KNN. Each model was evaluated ith training and testing data split of 70% and 30% respectively. From the performance metrics it has been observed that DT achieved maximum, accuracy of 99.76%, precision of 99.82%, recall of 99.75%, and F1 score of 99.78%. Then DT model successfully identified and classified the OC switch faults within one fundamental time period. All simulations were performed in the MATLAB/Simulink platform. The proposed methodology is adaptable and can be applied to detect and diagnose power device faults in various MLI configurations, which is helpful for applications like as PV, EV, and ID.

References

1. Poorfakhraei A, Narimani M, Emadi A (2021) A review of multilevel inverter topologies in electric vehicles: current status and future trends. IEEE Open J Power Electron 2:155–170. https://doi.org/10.1109/OJPEL.2021.3063550
2. Bana PR, Panda KP, Naayagi RT, Siano P, Panda G (2019) Recently developed reduced switch multilevel inverter for renewable energy integration and drives application: topologies, comprehensive analysis and comparative evaluation. IEEE Access 7:54888–54909. https://doi.org/10.1109/ACCESS.2019.2913447
3. Kumari A, Gopal Y, Dhaked DK, Panda KP, Vijaya Kumar YN (2023) A single source five-level switched-capacitor based multilevel inverter with reduced device count. e-Prime-Adv Electr Eng, Electron Energy 5:100235. ISSN 2772–6711
4. Gopal Y, Kumari A, Panda KP, Dhaked DK, Arya Y (2025) Implementation and analysis of switched-capacitor multilevel inverters in solar photovoltaic system. Sustain Energy Technol Assess 75:104227. ISSN 2213–1388. https://doi.org/10.1016/j.seta.2025.104227
5. Gupta KK, Ranjan A, Bhatnagar P, Sahu LK, Jain S (2016) Multilevel inverter topologies with reduced device count: a review. IEEE Trans Power Electron 31(1):135–151. https://doi.org/10.1109/TPEL.2015.2405012
6. Zhao S, Blaabjerg F, Wang H (2021) An overview of artificial intelligence applications for power electronics. IEEE Trans Power Electron 36(4):4633–4658. https://doi.org/10.1109/TPEL.2020.3024914
7. Aviña-Corral V, Rangel-Magdaleno JJ, Barron-Zambrano JH, Rosales-Nuñez S (2024) Review of fault detection techniques in power converters: fault analysis and diagnostic methodologies. Elsevier B.V. 234, Art. No. 114864. https://doi.org/10.1016/j.measurement.2024.114864
8. Yang W, Gu J, Xie X, Wei X, Ye H (2024) Lightweight shuffle–SimAM network-based open-circuit fault diagnosis of grid-connected cascaded H-bridge inverters. J Power Electron. https://doi.org/10.1007/s43236-024-00906-1
9. Hang J, Zhang C, Wang Y, Ding S, Zhao J (2024) Capacitor voltage change rate based fault localization strategy for modular multilevel converters. J Power Electron 24(3):361–372. https://doi.org/10.1007/s43236-023-00733-w
10. Anand A, Vinayak BA, Raj N, Jagadanand G, George S (2020) A generalized switch fault diagnosis for cascaded h-bridge multilevel inverters using mean voltage prediction. IEEE Trans Ind Appl 56(2):1563–1574. https://doi.org/10.1109/TIA.2019.2959540
11. Yang G, Hussain H, Li S, Zhang X, Yang J, Lee CHT (2023) Design and analysis of universal natural fault-tolerant SVPWM strategy with simplified fault diagnosis for multiphase motor drives. IEEE J Emerg Sel Top Power Electron 11(4):4340–4354. https://doi.org/10.1109/JESTPE.2023.3270428
12. Goel R, Etta D, Chavali RV, Dey A (2025) A fault tolerant modulation scheme for a hybrid multilevel inverter in the event of open-switch fault. IEEE Trans Power Electron 40(3):3861–3872. https://doi.org/10.1109/TPEL.2024.3486006
13. Kumar M (2022) Time-domain characterization and detection of open-circuit faults for the H-bridge power cell. IEEE Trans Power Electron 37(2):2152–2164. https://doi.org/10.1109/TPEL.2021.3103851
14. Ul Islam Khan M et al. (2024) Securing electric vehicle performance: machine learning-driven fault detection and classification. IEEE Access 12:71566–71584. https://doi.org/10.1109/ACCESS.2024.3400913
15. Stonier AA, Lehman B (2018) An intelligent-based fault-tolerant system for solar-fed cascaded multilevel inverters. IEEE Trans Energy Convers 33(3):1047–1057. https://doi.org/10.1109/TEC.2017.2786299
16. Jiang W, Wang C, Li Y-P, Wang M (2012) Fault detection and remedy of multilevel inverter based on BP neural network. In: 2012 Asia-Pacific power and energy engineering conference, Shanghai, China, pp 1–4. https://doi.org/10.1109/APPEEC.2012.6307658

17. Sivapriya A, Kalaiarasi N, Verma R, Chokkalingam B, Munda JL (2023) Fault diagnosis of cascaded multilevel inverter using multiscale kernel convolutional neural network. IEEE Access 11:79513–79530. https://doi.org/10.1109/ACCESS.2023.3299852
18. Ali M, Din Z, Solomin E, Cheema KM, Milyani AH, Che Z (2021) Open switch fault diagnosis of cascade H-bridge multi-level inverter in distributed power generators by machine learning algorithms. Energy Rep 7:8929–8942. https://doi.org/10.1016/j.egyr.2021.11.058
19. Hu X, Jia H, Zhang Y, Deng Y (2023) An open-circuit faults diagnosis method for MMC based on extreme gradient boosting. IEEE Trans Industr Electron 70(6):6239–6249. https://doi.org/10.1109/TIE.2022.3194584
20. Singh V, Yadav A, Gupta S, Abdelaziz AY (2024) Switch fault identification scheme based on machine learning algorithms for PV-Fed three-phase neutral point clamped inverter. e-Prime-Adv Electr Eng Electron Energy 8. https://doi.org/10.1016/j.prime.2024.100582
21. Gupta KK, Jain S (2014) A novel multilevel inverter based on switched DC sources. IEEE Trans Industr Electron 61(7):3269–3278. https://doi.org/10.1109/TIE.2013.2282606

Advancement in MMC Topologies and Balancing Techniques for Medium- and High-Power Applications

Jyoti Ranjan Dash, Rudranarayan Pradhan, and Pramod Agarwal

Abstract Modular Multilevel Converters (MMCs) have become a leading topology for medium- and high-voltage applications due to their modular structure, excellent scalability, and inherent fault-tolerant characteristics. The ability of MMCs to generate high-quality output waveforms with minimal filtering, combined with their flexible voltage and power scalability, makes them well suited for demanding applications such as High-Voltage Direct Current (HVDC) transmission, renewable energy grid integration, and industrial motor drives. This chapter investigates the implementation of Pulse Width Modulation (PWM) techniques across various Sub-Module (SM) configurations employed in MMCs, specifically Half-Bridge Sub-Modules (HBSMs), Full-Bridge Sub-Modules (FBSMs), and Flying Capacitor Sub-Modules (FCSMs). Each SM topology is examined in detail with respect to its switching methodology, operational characteristics, and capacitor voltage balancing requirements, offering a comprehensive analysis of their performance and challenges. Dedicated capacitor voltage balancing algorithms are designed for the HBSM, FBSM, and FCSM configurations, each tailored to address the unique operational constraints and control complexities associated with the respective SM structures. Comprehensive simulation studies are carried out in the MATLAB/Simulink environment to assess the performance and robustness of the proposed PWM schemes and balancing strategies. The simulation results demonstrate effective voltage regulation across a wide range of operating conditions, validating the reliability of each approach. Furthermore, the comparative analysis underscores the trade-offs and performance distinctions among the different SM topologies, offering critical insights for the optimal selection and control of sub-modules in practical MMC applications.

J. R. Dash · P. Agarwal
Department of Electrical Engineering, Indian Institute of Technology Roorkee, Roorkee, India
e-mail: jranjandash@ee.iitr.ac.in

P. Agarwal
e-mail: pramod.agarwal@ee.iitr.ac.in

R. Pradhan (✉)
School of Electrical Sciences, Odisha University of Technology and Research, Bhubaneswar, Odisha, India
e-mail: rudranarayan@outr.ac.in

Y. Gopal et al. (eds.), *Multilevel Inverter Technology for Renewable Energy Systems*, Green Energy and Technology, https://doi.org/10.1007/978-3-032-03589-9_10

Keywords Modular multilevel converter · Half-bridge sub-modules · Full-bridge sub-modules · Flying capacitor sub-modules · Pulse width modulation · Capacitor voltage balancing

1 Introduction

The Modular Multilevel Converter (MMC) represents a significant advancement in power electronics, offering numerous advantages over traditional two-level and three-level converter topologies. Originally developed to address the increasing demands of high-voltage and high-power applications, the MMC has gained widespread acceptance across a variety of sectors, including High-Voltage Direct Current (HVDC) transmission, renewable energy systems, electric vehicle propulsion, and industrial motor drives. Its unique architecture, composed of cascaded Sub-Modules (SMs) in each phase leg typically configured as either half-bridge, full-bridge, and flying capacitor circuits, enables the synthesis of output voltages with a large number of discrete levels [1, 2]. This results in a near-sinusoidal output waveform, significantly reducing Total Harmonic Distortion (THD) and removing the requirement of external filter.

One of the core strengths of the MMC lies in its modularity. The converter is easily scalable by adding or removing SMs, allowing for flexible voltage and power ratings. Each SM operates at a fraction of the total system voltage, which reduces the stress on power semiconductor devices and allows for the use of standard, widely available components. Moreover, the distributed storage of energy across multiple SMs contributes to improved fault tolerance, as faulty units can be bypassed or replaced without major disruption to the system [3]. The low-voltage steps between output levels also lead to lower dv/dt stress, reducing electromagnetic interference and enhancing the system's electromagnetic compatibility.

Despite the added complexity in control and balancing of SM capacitor voltages, the MMC offers substantial gains in efficiency, output quality, and reliability. These features collectively make the MMC a superior solution compared to conventional converter technologies, especially in applications where high performance, scalability, and power quality are critical. As the demand for efficient and flexible power conversion systems continues to grow, the MMC is poised to play a pivotal role in the future landscape of power electronics.

2 Modular Multilevel Converter

The MMC is a widely adopted multilevel topology, especially well suited for medium- and high-voltage applications. One of its key advantages is the ability to achieve high-voltage operation using low-voltage-rated switches. This is possible because the total DC-link voltage is divided among multiple SMs connected in series. As a result, each switch only experiences the relatively low voltage of its own SM capacitor, reducing the stress on individual components and allowing the use of cost-effective, lower-rated power devices. Furthermore, the output voltage of the MMC is formed by combining the voltages of selected SMs in a stepped manner. With a sufficient number of SMs, this stepped waveform closely approximates a sinusoidal shape, thereby minimizing harmonic distortion. This significantly reduces the need for bulky output filters and improves power quality, making the MMC a highly efficient and scalable solution for modern power conversion systems.

2.1 Operation Principle

The structural configuration of a three-phase MMC with a phase sequence of $(A - B - C)$ is depicted in Fig. 1. The converter is supplied by a split DC source, with each half providing $\frac{V_{dc}}{2}$. Each phase leg of the MMC comprises two arms, namely, the top arm represented as (t) and the bottom arm as (b) each consisting of (n) SMs and an arm inductor of inductance (L) and its internal resistance (r). The output of the converter is connected to a three-phase load, which typically exhibits both resistive and inductive characteristics. The MMC generates three-phase AC output

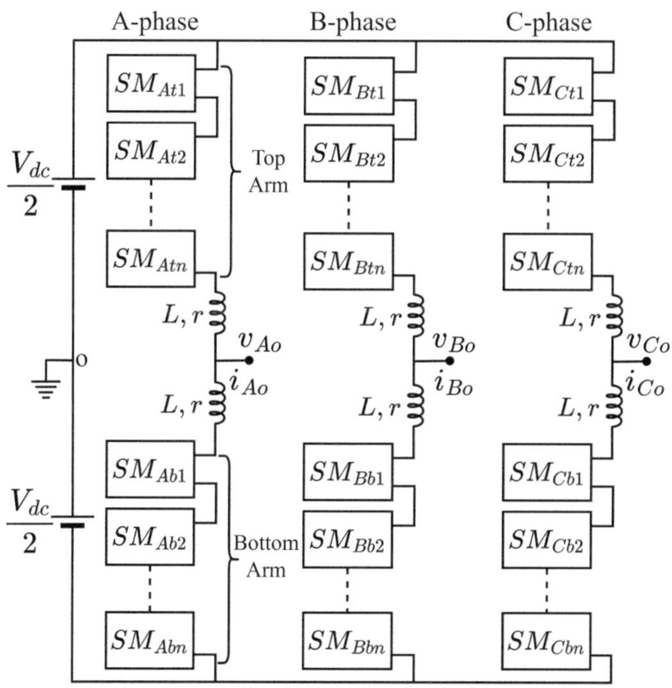

Fig. 1 Three-phase MMC

voltages (v_{Ao}, v_{Bo}, v_{Co}) and corresponding phase currents (i_{Ao}, i_{Bo}, i_{Co}), delivering high-quality sinusoidal waveforms [4, 5]. The major focus of this configuration is as follows:

- The MMC's modular architecture enables easy scaling of voltage and power ratings by adding or removing SMs as needed. This flexibility allows the converter to be adapted efficiently for various applications while simplifying maintenance and enhancing reliability.
- It can generate output voltage and current waveforms with low dv/dt and minimal ripple. In particular, the output voltage exhibits very low Total Harmonic Distortion (THD), ensuring high-quality power delivery.
- The MMC generates output voltage waveforms consisting of numerous voltage levels, allowing the SMs to switch at much lower frequencies.
- Redundant SMs can be incorporated in each arm to enable fault-tolerant operation.

2.2 *Fundamentals of Sub-modules in MMC*

The SM, as the fundamental building block of the MMC, plays a pivotal role in determining its operational performance. Each SM typically comprises a DC-link capacitor and a set of power semiconductor switches arranged in either a half-bridge or full-bridge, and many more [6]. These semiconductor devices enable the SM to either insert its stored capacitor voltage into the converter arm or bypass it, facilitating fine-grained modulation of the output voltage waveform. By connecting multiple SMs in series within each arm, the MMC synthesizes a high-fidelity multilevel voltage waveform that closely approximates a sinusoidal shape. The modular structure of SMs not only permits high-voltage operation using low-voltage-rated components but also enhances the MMC's scalability. The different types of SM topology are provided below:

- Half-Bridge Sub-Module (HBSM).
- Full-Bridge Sub-Module (FBSM).
- Flying Capacitor Sub-Module (FCSM).

2.2.1 Half-Bridge Sub-module

This study focuses on the widely adopted HBSM, which serves as a fundamental configuration in MMCs. Each HBSM comprises two Insulated Gate Bipolar Transistors (IGBTs), denoted as T_1 and $\overline{T}_1$, each equipped with an anti-parallel body diode, along with a DC-link electrolytic capacitor, as illustrated in Fig. 2. The IGBT switches used in the HBSM are unipolar in voltage-blocking capability and bidirectional in current conduction. By appropriately gating these switches, the HBSM can either insert the capacitor voltage v_c into the converter arm or bypass the capacitor. Consequently,

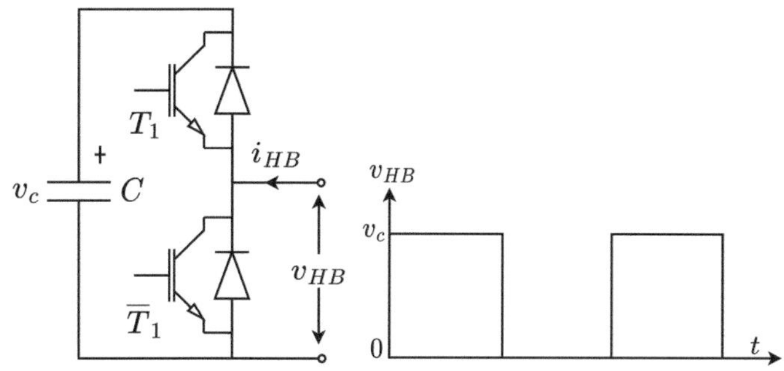

Fig. 2 Half-bridge sub-module

the HBSM is capable of producing two discrete output voltage levels: $v_{HB} = v_c$ or 0, thereby enabling the step-wise synthesis of the desired output waveform. When the arm current i_{HB} is positive, the capacitor is charged; conversely, when i_{HB} is negative, the capacitor discharges.

2.2.2 Full-Bridge Sub-module

The FBSM consists of two half-bridge circuits integrated with a shared electrolytic capacitor which is presented in Fig. 3. It offers several advantages over its half-bridge counterpart, making it particularly suitable for advanced MMC applications. One of the primary benefits of the FBSM is its capability to generate bipolar output voltages $v_{FB} = +v_c$, 0, and $-v_c$ which enables the converter to produce a more flexible and symmetrical output waveform. In addition, the FBSM provides inherent DC fault blocking capability, as it can interrupt the fault current path by inserting a negative voltage or fully isolating the faulted segment. This feature is especially valuable in applications such as HVDC transmission and grid-connected systems, where enhanced fault tolerance and system reliability are of paramount importance [7–9].

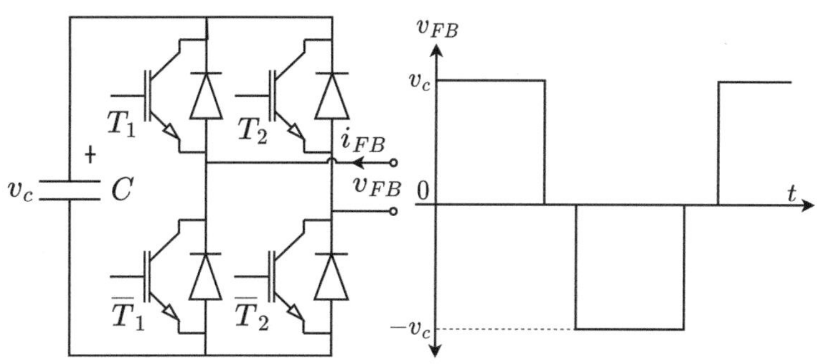

Fig. 3 Full-bridge sub-module

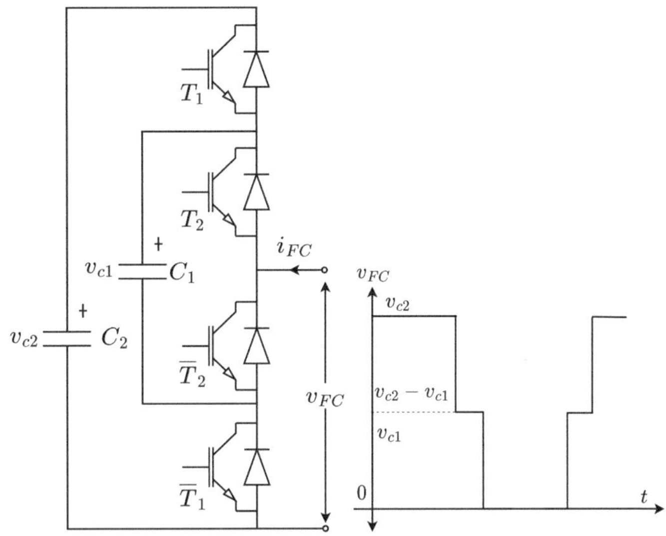

Fig. 4 Flying capacitor sub-module

2.2.3 Flying Capacitor Sub-module

The Flying Capacitor Sub-module (FCSM) enhances the performance of MMCs by effectively reducing voltage ripple and circulating currents, while simultaneously improving overall system efficiency. These advantages make the FCSM a compelling alternative to the conventional HBSM which is illustrated in Fig. 4. The FCSM topology comprises four IGBT switches and two flying capacitors with distinct capacitances, denoted as C_1 and C_2. This configuration enables the SM to generate three distinct output voltage levels: $V_{FC} = v_{c2}$, $V_{FC} = v_{c2} - v_{c1}$, or $V_{FC} = 0$. The ability to produce intermediate voltage levels contributes to a finer resolution in the synthesized output waveform, leading to improved voltage balancing and reduced harmonic distortion across the converter [10–14].

3 PWM Technique for Different Sub-modules

Pulse Width Modulation (PWM) is a key technique employed to regulate the AC output voltage in power electronic converters. For MMCs, PWM strategies are developed to reduce output voltage harmonics and improve the amplitude of the fundamental voltage at a given switching frequency. Beyond voltage control, these schemes are also formulated to meet additional control goals such as maintaining capacitor voltage balance, minimizing common-mode voltage, lowering switching frequency to reduce power losses, and decreasing output current ripple. Based on the operating switching frequency, PWM techniques for MMCs can be broadly categorized into high-frequency, low-frequency, and fundamental-frequency switching methods. The PWM techniques used for different SM uses in MMC are provided below:

3.1 Phase-Shifted Sine PWM for HBSM

In this study, a Phase-Shifted PWM (PSPWM) technique is employed. The PSPWM method used for HBSM utilizes n triangular carrier signals with a phase shift of $\phi_c = \frac{360^\circ}{n}$ for both the top and bottom arms of the converter. This modulation strategy allows the generation of $n + 1$ output voltage levels. To further enhance the number of output voltage levels, the phase difference between the carrier signals of the top and bottom arms is set to $\phi_{in} = \frac{360^\circ}{2n}$, commonly referred to as the interleaving angle. Due to this interleaving angle, the output phase voltage levels increase to $2n + 1$, improving waveform quality and system performance. To further illustrate the phase shifting technique, we consider an example with three SMs per arm $n = 3$. The carrier frequency used in the result is $f_c = 1$ kHz, while the sampling frequency is $f_s = 20$ kHz. The top arm carrier signals, bottom arm carrier signals, and reference waveform are shown in Fig. 5.

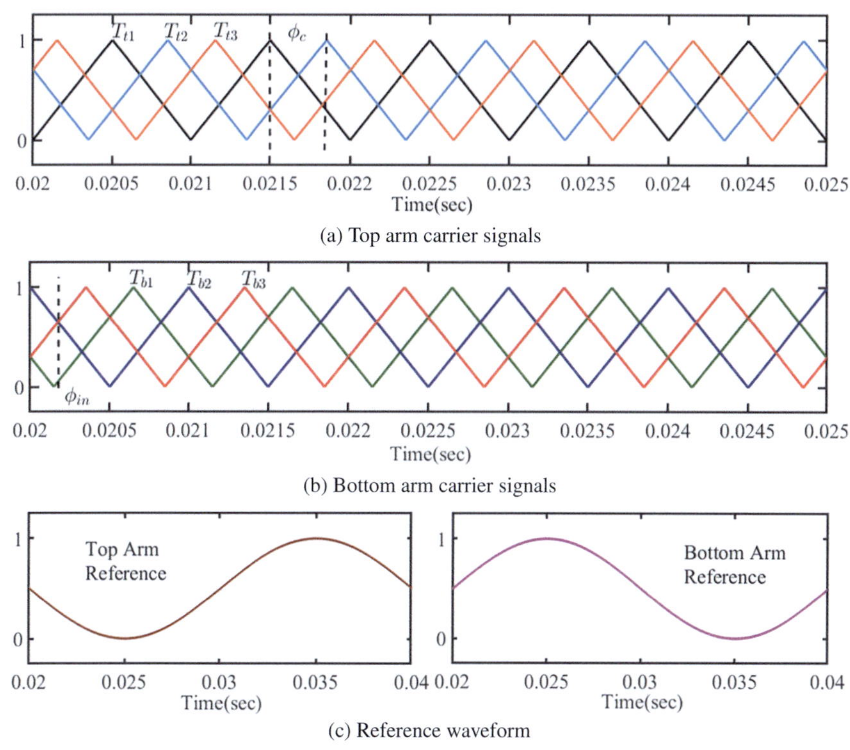

Fig. 5 PSPWM scheme for HBSM-based converter: **a** top arm carrier signals, **b** bottom arm carrier signals, and **c** reference waveform

3.2 Phase-Shifted Sine PWM for FBSM

In this study, a PSPWM technique is employed for the FBSM. When the number of SMs per arm is odd, the PSPWM scheme utilizes n triangular carrier signals, applied identically to both the top and bottom arms. In contrast, for an even number of SMs per arm, $2n$ triangular carrier signals are used n for the top arm and another n for the bottom arm. In both cases, the carrier signals are uniformly phase shifted by $\alpha_c = \frac{180^\circ}{n}$ within each arm. Furthermore, for even SM configurations, an additional phase shift between the carrier signals of the top and bottom arms is introduced, defined as $\alpha_{\text{in}} = \frac{180^\circ}{2n}$. This additional shift, commonly referred to as the interleaving angle, is not required for systems with an odd number of SMs. The interleaving strategy for even SM configurations increases the number of output phase voltage levels to $2n + 1$, thereby improving the waveform quality and enhancing the overall system performance.

As shown in Fig. 6, the system uses three SMs in each arm of the converter. Because of this configuration, introducing an interleaving phase shift between the carrier signals of the top and bottom arms is unnecessary. The modulation strategy employs two reference signals one assigned to the left HBSM and the other to the

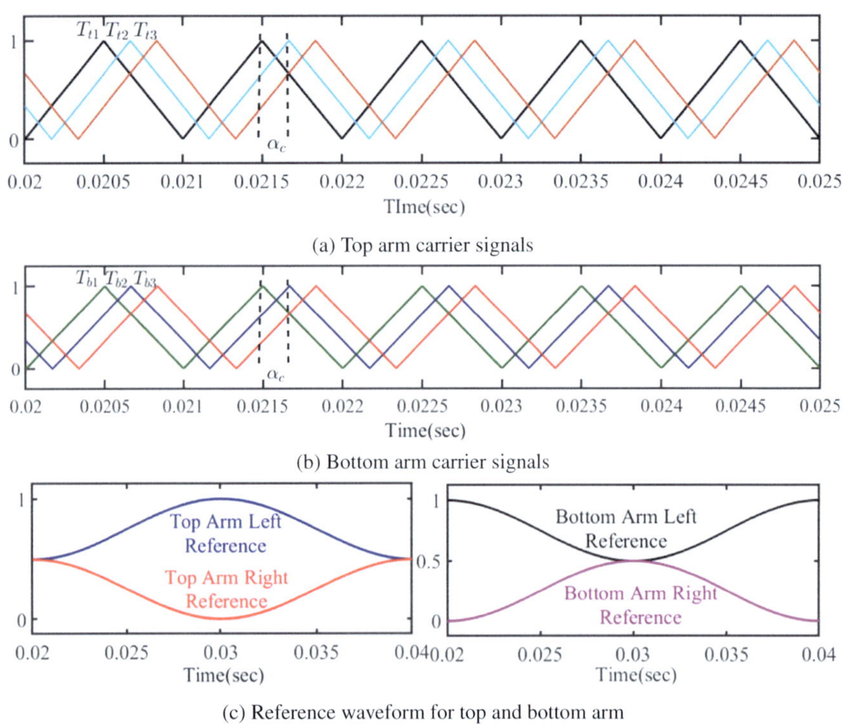

Fig. 6 PSPWM scheme for FBSM-based converter: **a** top arm carrier signals, **b** bottom arm carrier signals, and **c** reference waveform

right HBSM within the FBSM with a phase shift of 180°. The carrier frequency used for generating the PWM signals is $f_c = 1$ kHz, while the overall system operates with a sampling frequency of $f_s = 20$ kHz. The reference signals have a fundamental frequency of $f_l = 50$ Hz, corresponding to the desired output waveform frequency.

3.3 Phase-Shifted Sine PWM for FCSM

The PSPWM technique used in the FCSM generates an output voltage with $4n + 1$ levels. For an MMC with n SMs per arm, the modulation employs $2n$ triangular carrier signals n for the inner switches and n for the outer switches. A phase difference of 180° is introduced between the inner and outer carrier signals. Additionally, the phase difference between two consecutive inner or outer carrier signals is given by $\beta_c = \frac{180^\circ}{n}$. Additionally, a phase difference between the carrier signals of the top and bottom arms is introduced, defined as $\beta_{in} = \frac{180^\circ}{2n}$, which is commonly referred to as the interleaving angle. The reference and carrier signals corresponding to the FCSM are illustrated in Fig. 7.

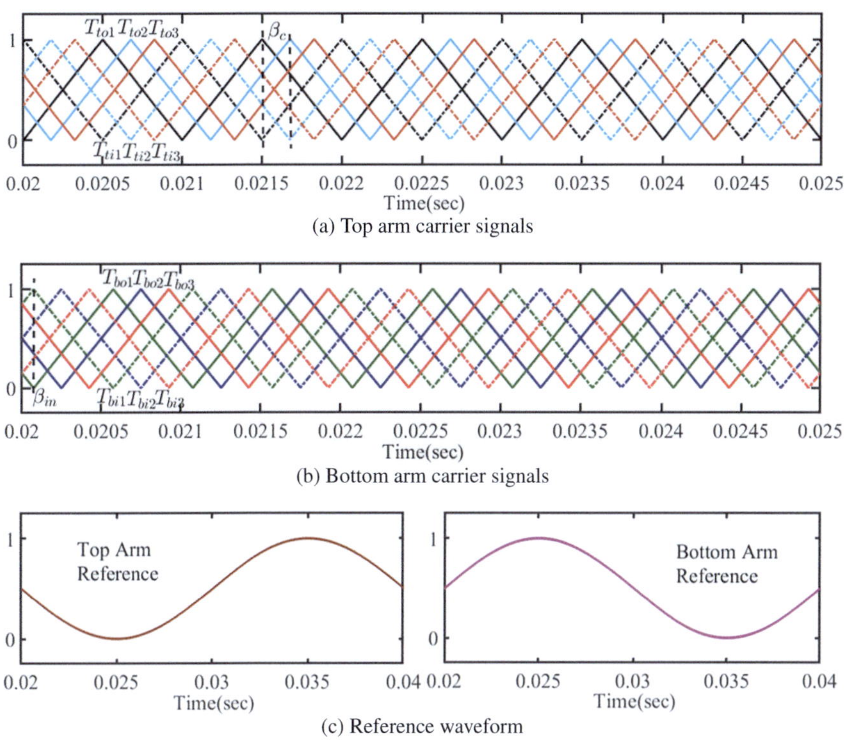

Fig. 7 PSPWM scheme for FCSM-based converter: **a** top arm carrier signals, **b** bottom arm carrier signals, and **c** reference waveform

4 Capacitor Voltage Balancing for Various Sub-module

Maintaining balanced capacitor voltages is critical for ensuring the stable and efficient operation of MMCs. Imbalances across SM capacitors can cause unequal voltage sharing, leading to power losses and degraded performance. Achieving uniform voltage levels enhances converter efficiency, improves power quality, and increases overall system dependability by avoiding over-voltage and under-voltage issues. Additionally, effective voltage balancing reduces the electrical stress on each capacitor, contributing to longer component life and greater system reliability [4, 5, 15, 16]. Capacitor voltage balancing for different SMs are presented below:

4.1 Capacitor Voltage Balancing for HBSM

Capacitor voltage balancing for the HBSM requires n voltage sensors, one arm current sensor, and access to the PWM signals for each arm. The shorting and selection algorithm-based voltage balancing method is compatible with various PWM schemes, providing enhanced flexibility in control implementation [5, 17]. The detailed capacitor voltage balancing methodology for the HBSM is presented in Algorithm 1, and the corresponding step-by-step execution procedure is outlined below:

- The top arm SM capacitor voltages are measured as $V_c = [v_{\text{tc}1}, v_{\text{tc}2}, \ldots, v_{\text{tc}n}]$, along with the top arm current i_t, and the sum of the gate signals, referred to as the desired number of active SMs or the level, denoted by LEVEL.
- It ensures that only a selected number of SMs determined by the required voltage level are turned on at each instant, while prioritizing capacitor voltage balancing. The algorithm takes as input the array of SM capacitor voltages V_c, the desired number of active SMs **LEVEL**, and the arm current direction i_t.
- If the arm current is positive, SMs with the lowest capacitor voltages are activated; if negative, those with the highest voltages are selected. A modified selection sort is used to sort the capacitor voltages and track corresponding indices for activation.
- If the level is zero, all SMs are turned off; if the level equals the total number of SMs, all are turned on. Otherwise, the algorithm selects the **LEVEL** number of SMs based on a sorting-like process.
- Finally, the gating signal vector **G** is updated to turn on the selected SMs.

4.2 Capacitor Voltage Balancing for FBSM

The FBSM consists of two HBSMs, arranged as the left and right halves, sharing a common capacitor. The proposed balancing algorithm generates two independent PWM signals—one for the left HBSM and another for the right HBSM. The left

Algorithm 1 Capacitor voltage balancing for HBSM based MMC

Require: Capacitor voltage array $V_c = [v_{\mathrm{tc}1}, v_{\mathrm{tc}2}, \ldots, v_{\mathrm{tc}n}]$;
Desired active submodule count *LEVEL*; Arm current direction i_{t}
Ensure: Gating signal array $\mathbf{G} = [g_1, g_2, \ldots, g_n]$
$n \leftarrow$ length of V_c
$\mathbf{G} \leftarrow \mathbf{0}_{1\times n}$ ▷ Initialize all gating signals to 0
if $LEVEL = 0$ **then**
    return G ▷ All submodules OFF
else if $LEVEL = n$ **then**
    $\mathbf{G} \leftarrow \mathbf{1}_{1\times n}$ ▷ All submodules ON
    return G
end if
indices $\leftarrow [1, 2, \ldots, n]$
for $i = 1$ to *LEVEL* **do**
    $minIdx \leftarrow i$
    for $j = i + 1$ to n **do**
        if $i_{\mathrm{t}} \geq 0$ **and** $V_c[j] < V_c[minIdx]$ **then**
            $minIdx \leftarrow j$
        else if $i_{\mathrm{t}} < 0$ **and** $V_c[j] > V_c[minIdx]$ **then**
            $minIdx \leftarrow j$
        end if
    end for
    if $minIdx \neq i$ **then**
        Swap $V_c[i] \leftrightarrow V_c[minIdx]$
        Swap $indices[i] \leftrightarrow indices[minIdx]$
    end if
end for
for $k = 1$ to *LEVEL* **do**
    $G[indices[k]] \leftarrow 1$
end for
return G

HBSM operates using the positive portion of the arm current, while the right HBSM operates with the negative portion. Depending on the switching level and direction of current flow, the capacitor charges or discharges accordingly, thereby maintaining voltage balance across the FBSM.The detailed capacitor voltage balancing algorithm for the FBSM is presented in Algorithm 2, and the step-by-step description is provided below:

- The algorithm begins by taking as input the capacitor voltage vector $V_c = [v_{\mathrm{tc}1}, \ldots, v_{\mathrm{tc}n}]$, the desired number of active SMs for both the left and right HBSMs ($LEVEL_1$ and $LEVEL_2$, respectively), and the arm current i_t. The goal is to generate two binary gating signal vectors $\mathbf{G}_1$ and $\mathbf{G}_2$, each of size n, which determine which SMs are activated in each half.
- At the core of this algorithm is a function named "SelectModules," which selects appropriate SMs for activation based on the capacitor voltage and the direction of current flow. The function first initializes the gating vector $\mathbf{G}$ as a zero vector, meaning all SMs are initially turned off.

Algorithm 2 PWM gating signals generation for top arms of FBSM MMC with n submodules

Require: Capacitor voltages $V_c = [v_{tc1}, \ldots, v_{tcn}]$;
Desired active submodule counts $LEVEL_1$, $LEVEL_2$; Arm current i_t
Ensure: Gating signals $\mathbf{G}_1, \mathbf{G}_2 \in \{0, 1\}^n$
function SELECTMODULES(V, $LEVEL$, I)
  $n \leftarrow$ length of V
  $\mathbf{G} \leftarrow \mathbf{0}_{1\times n}$
  if $LEVEL = 0$ **then**
    return G
  else if $LEVEL = n$ **then**
    $\mathbf{G} \leftarrow \mathbf{1}_{1\times n}$
    return G
  end if
  indices $\leftarrow [1, 2, \ldots, n]$
  for $i = 1$ to $LEVEL$ **do**
    $extremeIdx \leftarrow i$
    for $j = i + 1$ to n **do**
      if $(I \geq 0 \wedge V[j] < V[extremeIdx]) \vee (I < 0 \wedge V[j] > V[extremeIdx])$ **then**
        $extremeIdx \leftarrow j$
      end if
    end for
    if $extremeIdx \neq i$ **then**
      Swap $V[i] \leftrightarrow V[extremeIdx]$
      Swap $indices[i] \leftrightarrow indices[extremeIdx]$
    end if
  end for
  for $k = 1$ to $LEVEL$ **do**
    $\mathbf{G}[indices[k]] \leftarrow 1$
  end for
  return G
end function
$\mathbf{G}_1 \leftarrow$ SELECTMODULES(V_c, $LEVEL_1$, i_t)
$\mathbf{G}_2 \leftarrow$ SELECTMODULES(V_c, $LEVEL_2$, $-i_t$)
return $\mathbf{G}_1, \mathbf{G}_2$

- Next, the algorithm checks for two special cases. If the desired number of active SMs ("LEVEL") is zero, then no SM needs to be activated, and the zero vector is returned. If "LEVEL" equals the total number of SMs n, all SMs are activated, and the function returns a vector of ones.
- For general cases, the algorithm uses a sorting logic that depends on the direction of the arm current. If the current I is positive, it indicates that the SMs need to be charged, so the algorithm selects SMs with the lowest capacitor voltages first. Conversely, if the current is negative, the SMs are expected to discharge, so the ones with the highest voltages are prioritized. This is achieved using a selection-like sorting approach that rearranges the voltage vector and tracks the original indices accordingly.
- After sorting, the algorithm proceeds to activate the required number of SMs by setting the corresponding entries in the gating vector **G** to 1 based on the sorted

indices. This results in a gating signal that prioritizes either charging or discharging based on the current direction, while also maintaining capacitor voltage balance.
- Finally, the "SelectModules" function is called twice: once for the left HBSM using the arm current (i_t), and once for the right HBSM using the negative of the arm current ($-i_t$). This symmetric implementation ensures that the left and right halves of the FBSM balance the capacitor voltage effectively. The two resulting gating vectors, $\mathbf{G}_1$ and $\mathbf{G}_2$, are then returned as the output of the algorithm.

4.3 Capacitor Voltage Balancing Used in FCSM

The FCSM comprises four IGBT switches and two capacitors, namely, the inner and outer capacitors. In this topology, the outer capacitor is equal to the capacitance of the inner capacitor. To ensure appropriate voltage distribution across the capacitors, a single-capacitor voltage balancing algorithm is introduced. This algorithm regulates the capacitor voltages such that the voltage across the outer capacitor consistently remains twice that of the inner capacitor. Maintaining this proportional voltage relationship is essential for ensuring reliable SM operation and for minimizing voltage imbalance across the switching devices. The detailed capacitor voltage balancing methodology for the FCSM is presented in Algorithm 3, and the corresponding step-by-step execution procedure is outlined below:

- The algorithm begins by initializing the required inputs, which include the outer capacitor voltage vector $[v_{\text{tco}1}/2, \ldots, v_{\text{tco}n}/2]$, inner capacitor voltage vector $[v_{\text{tci}1}, \ldots, v_{\text{tci}n}]$, the desired number of ON SMs represented by LEVEL, and the corresponding arm current i_{t}. These inputs serve as the foundation for selecting appropriate gating signals to ensure balanced capacitor voltages across the FCSMs.
- To simplify the selection process and allow uniform treatment of both outer and inner capacitors, the algorithm combines the outer and inner capacitor voltage vectors into a single voltage vector V_c of dimension $2n$. This combined vector represents the voltage status of each of the $2n$ capacitors (outer and inner) distributed across the n SMs.
- Following this, a gating signal vector $\mathbf{G} \in \{0, 1\}^{2n}$ is initialized to all zeros. Each element of this vector corresponds to a specific switch in the FCSMs, and the goal of the algorithm is to appropriately update this vector based on the capacitor voltages and current direction so as to turn ON the optimal subset of switches.
- If the desired number of ON SMs is zero (LEVEL $= 0$), the algorithm terminates early by returning the all-zero gating signal vector $\mathbf{G}$, signifying that all SMs should remain OFF. This is a straightforward boundary condition to avoid unnecessary processing.
- In contrast, if LEVEL $= 2n$, meaning all switches across all SMs must be activated, the algorithm updates all entries of $\mathbf{G}$ to one and returns this vector. This ensures that all capacitor paths are engaged when full SM operation is required.

Algorithm 3 Gating Signal Selection for FCSM Based MMC with n Submodules

```
Require: Outer capacitor voltages [v_tco1, ..., v_tcon];
    Inner capacitor voltages [v_tci1, ..., v_tcin];
    Desired number of ON submodules LEVEL;
    Arm current i_t
Ensure: Gating signal vector G = [g_1, g_2, ..., g_2n]
V_c ← [v_tco1/2, v_tci1, ..., v_tcon/2, v_tcin]
G ← 0_{1×2n}                                  ▷ Initialize all gating signals to 0
if LEVEL = 0 then
    return G                                  ▷ All submodules OFF
else if LEVEL = 2n then
    G ← 1_{1×2n}                              ▷ All submodules ON
    return G
end if
indices ← [1, 2, ..., 2n]
for i = 1 to LEVEL do
    targetIdx ← i
    for j = i + 1 to 2n do
        if i_t ≥ 0 and V_c[j] < V_c[targetIdx] then
            targetIdx ← j
        else if i_t < 0 and V_c[j] > V_c[targetIdx] then
            targetIdx ← j
        end if
    end for
    if targetIdx ≠ i then
        Swap V_c[i] ↔ V_c[targetIdx]
        Swap indices[i] ↔ indices[targetIdx]
    end if
end for
for k = 1 to LEVEL do
    G[indices[k]] ← 1
end for
return G
```

- For partial-level switching, the algorithm proceeds to create an index vector indices $= [1, 2, \ldots, 2n]$, which maintains the original positions of the capacitor voltages in the combined vector V_c. This index tracking is essential when performing sorting and rearrangement of voltage values.
- The core logic involves a selective sorting mechanism based on the direction of arm current i_t. If the arm current is positive, it indicates capacitor charging, and hence SMs with the lowest voltages are prioritized. Conversely, if the current is negative, indicating discharging, the SMs with the highest voltages are preferred. This directional strategy ensures that capacitor voltages are balanced by either selectively charging or discharging them.
- To facilitate this selection, the algorithm iteratively compares capacitor voltages and identifies the most suitable SM index (targetIdx) within the current pass. If a better candidate is found, the algorithm swaps the voltage values in V_c and their corresponding indices in indices. This step ensures that the top LEVEL SMs with appropriate voltage characteristics are sorted toward the beginning of the vector.

- After identifying the desired SMs, the algorithm updates the corresponding positions in the gating signal vector **G**, as indicated by the sorted indices, by setting those elements to 1. This final assignment marks the selected switches to be turned ON in the current switching cycle.
- Finally, the completed gating signal vector **G** is returned. This vector is then used to drive the IGBT switches within each FCSM to achieve balanced capacitor voltage operation, thereby ensuring stable and efficient performance of the modular multilevel converter.

5 Balancing Results for Various Sub-module Types

To facilitate a better understanding of different SM capacitor voltage balancing techniques, a MATLAB/Simulink-based simulation study has been conducted. Figure 8a illustrates the single-phase configuration of an MMC used for this analysis. Three different SM topologies are considered for comparison: the HBSM, as shown in Fig. 8b; the FBSM, depicted in Fig. 8c; and the FCSM, presented in Fig. 8d. Each topology is implemented within the same MMC framework to evaluate and compare the performance of capacitor voltage balancing algorithms under identical operating conditions.

The simulation is carried out using the following parameters: the input DC voltage is set to $V_{dc} = 6000$ V, with a sampling frequency of $f_s = 40$ kHz, a carrier frequency of $f_c = 1$ kHz, and a line frequency of $f_l = 50$ Hz. The modulation index is maintained at $m = 1$, and the number of SMs in each arm is $n = 3$. For the SM parameters, each SM is equipped with a capacitance of $C = 4700\,\mu$F, while the arm inductance and

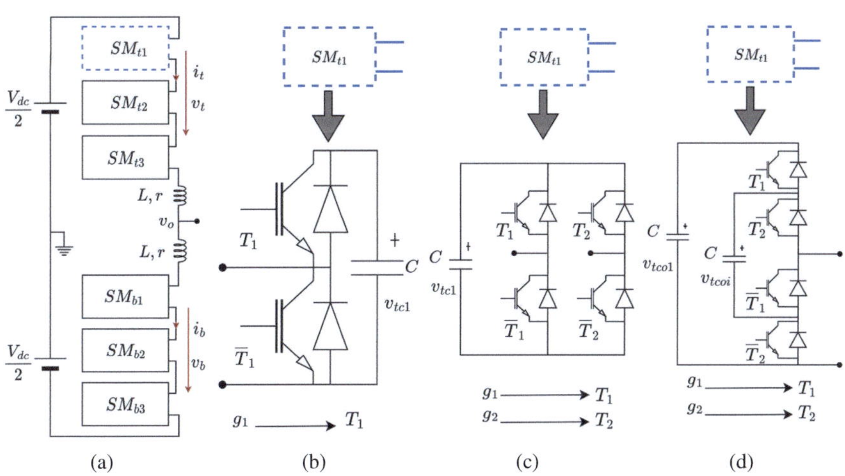

Fig. 8 Single-phase three SM in each arm structure of MMC **a** single-phase MMC, **b** HBSM, **c** FBSM, and **d** FCSM

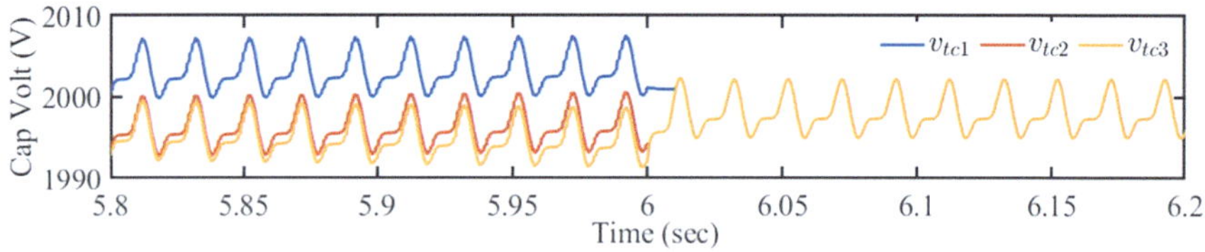

(a) Top arm SM capacitor voltages

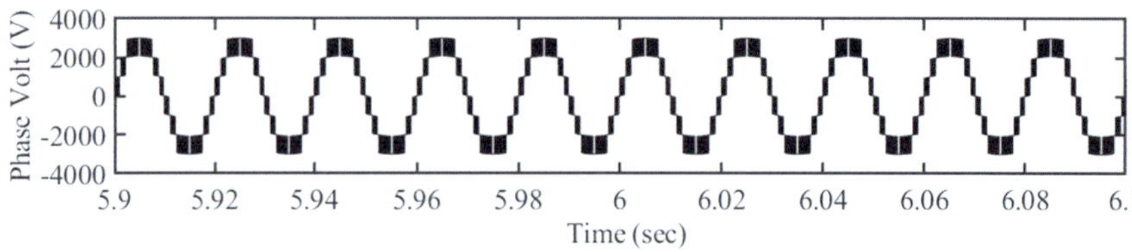

(b) Phase voltage

Fig. 9 HBSM voltage waveform **a** top arm SM capacitor voltage both in unbalance and balance conditions (balancing applied at 6 s) **b** output phase voltage

internal resistance are $L = 2\,\text{mH}$ and $r = 0.4\,\Omega$, respectively. A resistive-inductive (R-L) load is considered in the simulation, comprising a resistance $R_o = 100\,\Omega$ and an inductance $L_o = 20\,\text{mH}$. These parameters provide a realistic operating scenario for evaluating the performance of different SM topologies and their corresponding balancing strategies.

In the first case, the HBSM topology is considered for simulation. Figure 9 presents the capacitor voltage and output voltage waveforms of the HBSM-based MMC. As shown in Fig. 9a, the capacitor voltages are initially unbalanced and balancing applied at $t = 6\,\text{s}$. The corresponding output voltage during the unbalanced and balance conditions is depicted in Fig. 9b. In this configuration, three HBSMs are employed in each arm, resulting in a seven-level output voltage waveform.

In the second case, the FBSM topology is employed for simulation. Figure 10 illustrates the capacitor voltage and output voltage waveforms of the FBSM-based MMC. As depicted in Fig. 10a, the capacitor voltages are initially unbalanced, and voltage balancing is initiated at $t = 6$ s. The corresponding output voltage waveforms, under both unbalanced and balanced conditions, are shown in Fig. 10b. Similar to the previous case, three FBSMs are used in each arm, yielding a seven-level output voltage.

In the third case, the FCSM topology is utilized for simulation. Figure 11 shows the capacitor voltage and output voltage waveforms of the FCSM-based MMC. As illustrated in Fig. 11a and b, the outer and inner capacitor voltages are initially unbalanced, with voltage balancing applied at $t = 6$ s. Although both capacitors have equal capacitance, the outer capacitor voltage is maintained at twice the voltage of the inner capacitor. The corresponding output voltage waveforms during the unbalanced and balanced conditions are shown in Fig. 11c. In this setup, three FCSMs are implemented in each arm, resulting in a 13-level output voltage waveform.

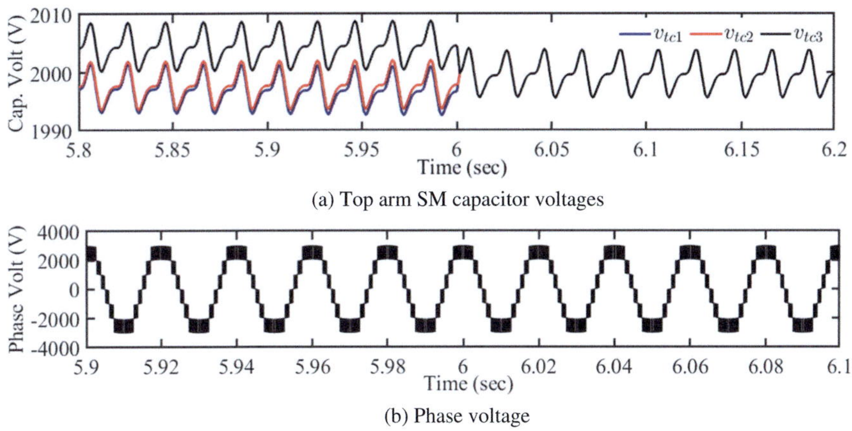

Fig. 10 FBSM voltage waveform **a** top arm SM capacitor voltage both in unbalance and balance conditions (balancing applied at 6 s) **b** output phase voltage

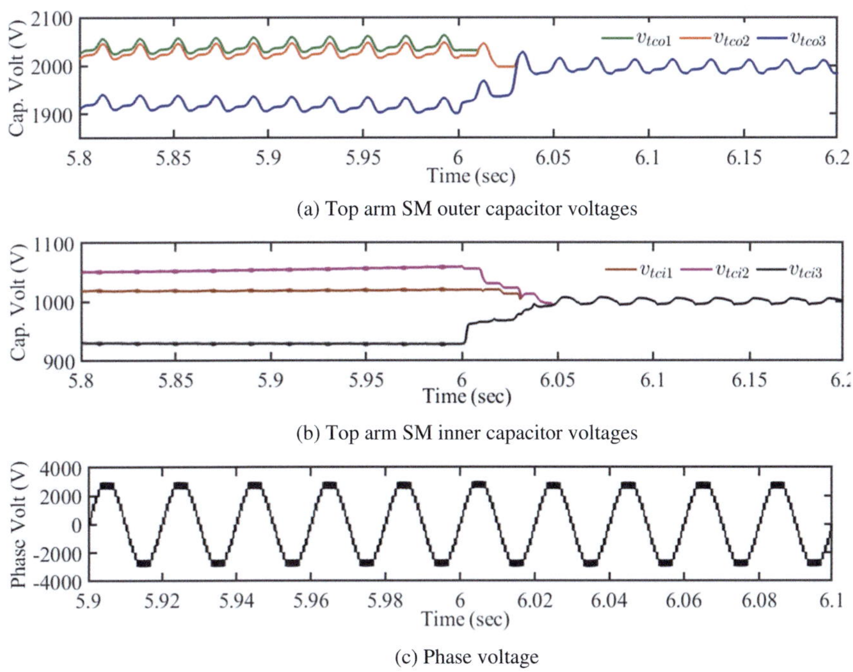

Fig. 11 FCSM voltage waveform **a** top arm SM outer capacitor voltage both in unbalance and balance conditions (balancing applied at 6 sec), **b** inner capacitor voltage, **c** output phase voltage

6 Conclusions

This chapter has presented a comprehensive analysis of capacitor voltage balancing strategies in Modular Multilevel Converters (MMC) for three prominent Sub-Module (SM) topologies: half-bridge, full-bridge, and flying capacitor. The discussion focused on the unique characteristics of each topology and the control methods employed to achieve effective voltage balancing under various operating conditions. Each topology was evaluated using a Phase-Shifted Pulse Width Modulation (PSPWM) scheme. The capacitor voltage balancing algorithms based on sorting and selection techniques are discussed in detail.

The simulation results demonstrated that all three SM configurations were capable of achieving effective capacitor voltage balancing when appropriate control algorithms were activated. For the HBSM and FBSM topologies, voltage balancing was successfully established after the activation of the control strategy at $t = 6$ s, resulting in stable capacitor voltages and a well-defined seven-level output voltage waveform. In the case of the FCSM topology, which features both inner and outer capacitors, the balancing algorithm effectively maintained the outer capacitor voltage at twice that of the inner capacitor. This regulation led to the generation of a high-resolution 13-level output waveform. These results underscore the superior voltage quality provided by the FCSM topology, albeit with the trade-off of increased control complexity.

The comparative evaluation highlights that while HBSMs and FBSMs offer simpler control schemes and satisfactory output performance, FCSMs provide superior waveform quality and improved capacitor voltage utilization. However, this advantage comes at the cost of increased control complexity due to the need for more advanced balancing techniques. These insights are crucial for informing the optimal selection of SM topologies based on application-specific requirements in MMC-based high-voltage and high-performance power electronic systems.

References

1. Debnath S, Qin J, Bahrani B, Saeedifard M, Barbosa P (2015) Operation, control, and applications of the modular multilevel converter: a review. IEEE Trans Power Electron 30(1):37–53
2. Perez MA, Ceballos S, Konstantinou G, Pou J, Aguilera RP (2021) Modular multilevel converters: recent achievements and challenges. IEEE Open J Ind Electron Soc 2:224–239
3. He J, Yang Q, Wang Z (2020) On-line fault diagnosis and fault-tolerant operation of modular multilevel converters—a comprehensive review. CES Trans Electr Mach Syst 4(4):360–372
4. Dash JR, Agarwal P (2022) Voltage balancing in mmc operated at reduced switching frequency without using sensors. In: IEEE IAS global conference on emerging technologies (GlobConET), pp 545–550
5. Dash JR, Mohanty PK, Agarwal P, Jena P, Padhy NP (2025) Temporal convolutional network-based capacitor voltage prediction with reduced switching frequency for voltage balancing in MMC. IEEE Trans Ind Appl:1–15
6. Hasan NS, Rosmin N, Osman DAA, Musta'amal@Jamal AH (2017) Reviews on multilevel converter and modulation techniques. Renew Sustain Energy Rev 80:163–174

7. Li B, Yang R, Xu D, Wang G, Wang W, Xu D (2015) Analysis of the phase-shifted carrier modulation for modular multilevel converters. IEEE Trans Power Electron 30(1):297–310
8. Thakur SS, Odavic M, Allu A, Zhu ZQ, Atallah K (2022) Analytical modelling and optimization of output voltage harmonic spectra of full-bridge modular multilevel converters in boost mode. IEEE Trans Power Electron 37(3):3403–3420
9. Yin J, Dai N, Vazquez S, Perez MA, Zhang B, Leon JI, Franquelo LG (2025) Direct pulsewidth modulation technique for modular multilevel converters based on full-bridge submodules. IEEE Trans Power Electron 40(1):1085–1098
10. Dekka A, Wu B, Zargari NR (2017) Start-up operation of a modular multilevel converter with flying capacitor submodules. IEEE Trans Power Electron 32(8):5873–5877
11. Dekka A, Wu B, Zargari NR (2015) A novel modulation scheme and voltage balancing algorithm for modular multilevel converter. In: IEEE applied power electronics conference and exposition (APEC), pp 1227–1233
12. Adam GP, Abdelsalam I, Fletcher JE, Burt GM, Holliday D, Finney SJ (2017) New efficient submodule for a modular multilevel converter in multiterminal HVDC networks. IEEE Trans Power Electron 32(6):4258–4278
13. Dekka A, Wu B, Zargari NR, Fuentes RL (2016) Dynamic voltage balancing algorithm for modular multilevel converter: a unique solution. IEEE Trans Power Electron 31(2):952–963
14. Dekka A, Wu B, Yaramasu V, Zargari NR (2017) Model predictive control with common-mode voltage injection for modular multilevel converter. IEEE Trans Power Electron 32(3):1767–1778
15. Samajdar D, Bhattacharya T, Dey S (2019) A reduced switching frequency sorting algorithm for modular multilevel converter with circulating current suppression feature. IEEE Trans Power Electron 34(11):10480–10491
16. Dash JR, Mohanty PK, Agarwal P, Jena P, Padhy NP (2023) A deep LSTM-based RSF capacitor voltage balancing method for MMC. In: IEEE 3rd international conference on sustainable energy and future electric transportation (SEFET), pp 1–6
17. Choudhury A, Shimada T, Kanouda A, Mabuchi Y (2020) Reduced switching loss based cell capacitor voltage balancing strategy for MMC. In: IEEE international conference on power electronics, drives and energy systems (PEDES), pp 1–5

Index

A
Anti-predatory Particle Swarm Optimization (APSO), 87, 89, 98–102, 104
Artificial neural networks, 61, 63–70, 72, 90, 144

B
Boosting Common-Ground Multilevel Inverter (BCGMLI), 107, 109, 110

C
Capacitor voltage balancing, 56, 157, 166, 167, 169, 171, 174
Common ground, 47, 50, 52, 54–57, 107–110, 117
Common-mode voltage, 45, 46, 50, 54, 110, 162
Cost function, 84
CT-type, 1, 3–6, 9, 12, 13, 14, 18

D
Dual 2-L 5-phase inverter, 23, 24, 27, 30, 34

F
Flying capacitor sub-modules, 157, 160, 162
Full-bridge sub-modules, 157, 160, 161
Fuzzy Logic Controller (FLC), 87, 88, 90–95, 104

H
Half-bridge sub-modules, 157, 160, 161

L
Leakage current, 45–47, 49–51, 54–57, 107, 109, 110, 117

M
MATLAB/SIMULINK model, 21, 30, 32, 135
Modular multilevel converter, 144, 145, 157–163, 165, 166, 167, 168, 170–172, 174
Multilevel inverters, 1–4, 6, 9, 13, 14, 18, 21, 22, 43, 45–47, 49, 50, 52, 54, 55, 57, 63, 66–69, 72, 75, 76, 87–89, 102, 104, 107–110, 113, 117, 122–130, 133–140, 143–147, 149, 151, 153

N
Nine-level inverter, 108

O
Open Circuit Fault (OCF), 122–125, 129, 133–138, 145, 147, 149, 150, 153
Open End Winding (OEW), 21–24, 34

P
Pulse width modulation, 1, 2, 9, 11, 21–23, 34, 50, 54, 57, 89, 103, 107, 110,

Y. Gopal et al. (eds.), *Multilevel Inverter Technology for Renewable Energy Systems*, Green Energy and Technology, https://doi.org/10.1007/978-3-032-03589-9

114, 121, 130, 138, 145, 157, 162–166, 168, 174
PV Renewable Energy (RE), 87

R
Reduce Device (RD), 121, 122, 138
Reduced device count, 1–3, 18, 75, 76, 85
Reliability, 2, 43, 45, 47, 50, 53, 54, 122, 139, 140, 144, 157, 158, 160, 161, 166
Renewable energy systems, 18, 55, 61, 63, 144, 158

S
Scaled conjugate gradient method, 61, 72
7L-PV-SCMLI topology, 87, 102
Solar PV system, 64, 66, 72, 88
Space Vector Pulse Width Modulation (SVPWM), 9, 21–24, 27, 29, 30, 32, 34, 35, 37, 38, 42
Switched capacitor, 2, 45–47, 50–55, 57, 61–64, 68, 69, 72, 75, 76, 82, 87, 96, 104, 108, 110, 112–115, 144, 147
Switched-capacitor cell, 108
Switched-capacitor MLI, 63, 68

T
Total Harmonic Distortion (THD), 2, 4, 12, 17, 21–23, 27, 29, 34–37, 39, 41–43, 45–47, 52, 55, 56, 64, 72, 87, 89, 96, 98–100, 102, 104, 118, 135, 138, 143, 144, 148, 150, 153, 158, 160
Total standing voltage, 1, 9, 14, 15, 18, 52, 53, 56, 75, 76, 84, 85, 103, 104, 109
Transformer-based MLIs, 45–47, 49
Transformerless inverter, 46, 108
Transformer-less MLIs, 50, 55

V
Voltage boost, 46, 51, 56, 57, 62, 64, 75, 76, 108, 109, 115

MIX
Papier aus verantwortungsvollen Quellen
Paper from responsible sources
FSC® C105338

If you have any concerns about our products,
you can contact us on
ProductSafety@springernature.com

In case Publisher is established outside the EU,
the EU authorized representative is:
Springer Nature Customer Service Center GmbH
Europaplatz 3, 69115 Heidelberg, Germany

Printed by Libri Plureos GmbH
in Hamburg, Germany